CALCULUS
UNLIMITED

CALCULUS
UNLIMITED

Jerrold Marsden
University of California, Berkeley

Alan Weinstein
University of California, Berkeley

The Benjamin/Cummings Publishing Company, Inc.
Menlo Park, California • Reading, Massachusetts
London • Amsterdam • Don Mills, Ontario • Sydney

Sponsoring Editor: Susan A. Newman
Production Editor: Madeleine S. Dreyfack

To Alison, Christopher, Asha, and the future of calculus.

Library of Congress Cataloging in Publication Data

Marsden, Jerrold E.
    Calculus unlimited.

    Includes index.
    1. Calculus. I. Weinstein, Alan, 1943-   joint
author. II. Title.
QA303.M3374   515   80-24446
ISBN 0-8053-6932-5

ABCDEFGHIJ-AL-83210

The Benjamin/Cummings Publishing Company, Inc.
2727 Sand Hill Road
Menlo Park, California 94025

Preface

Purpose

This book is intended to supplement our text, *Calculus* (Benjamin/Cummings, 1980), or virtually any other calculus text (see page vii, How To Use This Book With Your Calculus Text). As the title *Calculus Unlimited* implies, this text presents an alternative treatment of calculus using the method of exhaustion for the derivative and integral in place of limits. With the aid of this method, a definition of the derivative may be introduced in the first lecture of a calculus course for students who are familiar with functions and graphs.

Approach

Assuming an intuitive understanding of real numbers, we begin in Chapter 1 with the definition of the derivative. The axioms for real numbers are presented only when needed, in the discussion of continuity. Apart from this, the development is rigorous and contains complete proofs.

As you will note, this text has a more geometric flavor than the usual analytic treatment of calculus. For example, our definition of completeness is in terms of convexity rather than least upper bounds, and Dedekind cuts are replaced by the notion of a transition point.

Who Should Use This Book

This book is for calculus instructors and students interested in trying an alternative to limits. The prerequisites are a knowledge of functions, graphs, high school algebra and trigonometry.

How To Use This Book

Because the "learning-by-doing" technique introduced in *Calculus* has proved to be successful, we have adapted the same format for this book. The solutions to "Solved Exercises" are provided at the back of the book; however readers are encouraged to try solving each example before looking up the solution.

The Origin Of The Definition Of The Derivative

Several years ago while reading *Geometry and the Imagination*, by Hilbert and Cohn-Vossen (Chelsea, 1952, p. 176), we noticed a definition of the circle of curvature for a plane curve C. No calculus, as such, was used in this definition. This suggested that the same concept could be used to define the tangent line and thus serve as a limit-free foundation for the differential calculus. We introduced this new definition of the derivative into our class notes and developed it in our calculus classes for several years. As far as we know, the definition has not appeared elsewhere. If our presumption of originality is ill-founded, we welcome your comments.

Jerrold Marsden
Alan Weinstein
Berkeley, CA

How To Use This Book With Your Calculus Text

There are two ways to use this book:

1. It can be used to take a second look at calculus from a fresh point of view after completion of a standard course.

2. It can be used simultaneously with your standard calculus text as a supplement. Since this book is theory oriented, it is meant for better students, although Chapter 1 is designed to be accessible to all students.

The table below shows the chapters of this book that can be used to supplement sections in some of the standard calculus texts.

Chapters in this Book	Corresponding Chapters in Standard Texts*									
	Ellis/ Gulick	Grossman	Leithold	Marsden/ Weinstein	Protter/ Morrey	Purcell	Salas/ Hille	Shenk	Swokowski	Thomas/ Finney
1	3.2	2.10	3.3	R-3	4.3	4.3	3.1	3.1	3.2	1.8
2	3.2	2.10	3.3	1.1	4.3	4.3	3.1	3.2	3.2	1.9
3	3.1	3.2	3.5	1.3	5.1	5.2	3.2	3.4	3.3	2.3
4	1.1	1.2	1.1	R-1	A-4	1.3	1.5	1.2	1.1	1.2
5	2.5	2.4/10.4	2.5/5.1	2.3	6.5	6.3	2.6/App. B	2.7/4.2	2.5/4.4	2.11/3.1
6	4.6	3.9	5.2-5.5	2.4	6.4	6.12	4.7	4.3	4.5	3.4
7	4.2	2.4/10.6	4.7	3.3	6.2	6.2	4.1	4.12	4.3	3.8
8	8.2	3.7/10.7	9.3/3.6	5.2	9.4/5.2	5.11/10.2	3.7	4.6	8.7	2.4
9	3.3	7.5	10.2	6.2	9.2	10.7	7.1	3.8	9.2	2.10
10	8.4	6.3	9.2-9.5	7.2	9.9	10.3	6.4	7.5	8.4	6.8
11	5.3	4.3	7.3	4.1	7.3	8.3	5.2	5.1	5.2	4.8
12	5.4	4.5	7.6	4.2	7.5	8.4	5.3	5.2	5.5	4.8
13	6.5	10.8	3.4	12.2	4.6	8.4	App. B	App. Ch. 5	App. II	Ch. 4/Summary

*See following list of references.

- R. Ellis and D. Gulick, *Calculus with Analytic Geometry* Harcourt, Brace, Jovanovich (1978)

- S. Grossman, *Calculus* Academic Press (1977)

- L. Leithold, *Calculus*, 3rd Ed., Harper and Row (1976)

- J. Marsden and A. Weinstein, *Calculus*, Benjamin/Cummings (1980)

- M. Protter and C. B. Morrey, *Calculus*, 4th Ed., Addison-Wesley (1980)

- E. J. Purcell *Calculus with Analytic Geometry*, 3rd Ed., Prentice-Hall (1978)

- S. L. Salas and E. Hille, *Calculus. One and Several Variables*, 3rd Ed., Wiley (1978)

- A. Shenk, *Calculus*, Goodyear (1977)

- E. W. Swokowski, *Calculus with Analytic Geometry*, Second Edition, Prindle Weber and Schmidt, (1980)

- G. B. Thomas and R. L. Finney, *Calculus and Analytic Geometry*, 5th Ed., Addison-Wesley (1979)

Preview For The Instructor

This preview is intended for those who already know calculus. Others should proceed directly to Chapter 1.

The method of exhaustion of Eudoxus and Archimedes may be summarized as follows: Having defined and computed areas of polygons, one determines the area of a curvilinear figure F using the principle that whenever P_1 and P_2 are polygons such that P_1 is inside F and F is inside P_2, then Area $(P_1) \leq$ Area $(F) \leq$ Area (P_2). This approach appears in modern mathematics in the form of Dedekind cuts, inner and outer measure, and lower and upper sums for integrals.

To apply the method of exhaustion to differentiation, we replace the relation of inclusion between figures by the relation of *overtaking* defined as follows.

Definition Let f and g be real-valued functions with domains contained in $\mathbb{R}$, and x_0 a real number. We say that f *overtakes* g *at* x_0 if there is an open interval I containing x_0 such that

0. $x \in I$ and $x \neq x_0$ implies x is in the domain of f and g.

1. $x \in I$ and $x < x_0$ implies $f(x) < g(x)$.

2. $x \in I$ and $x > x_0$ implies $f(x) > g(x)$.

Given a function f and a number x_0 in its domain, we may compare f with the linear functions $l_m(x) = f(x_0) + m(x - x_0)$.

Definition Let f be a function defined in an open interval containing x_0. Suppose that there is a number m_0 such that:

1. $m < m_0$ implies f overtakes l_m at x_0.

2. $m > m_0$ implies l_m overtakes f at x_0.

Then we say that f *is differentiable at* x_0, and that m_0 is *the derivative* of f at x_0.

The following notion of *transition* occurs implicitly in both of the preceding definitions.

Definition   Let A and B be sets of real numbers, and x_0 a real number. We say that x_0 is a *transition point from A to B* if there is an open interval I containing x_0 such that:

1. $x \in I$ and $x < x_0$ implies $x \in A$ and $x \notin B$.
2. $x \in I$ and $x > x_0$ implies $x \in B$ and $x \notin A$.

The preceding definition of the derivative is equivalent to the usual limit approach, as we shall prove in Chapter 13. However, it is conceptually quite different, and for students who wish a logically complete definition of the derivative, we believe that it is simpler and geometrically appealing.

Both Euclid and Archimedes probably employed the following definition of tangent: "the tangent line touches the curve, and in the space between the line and curve, no other straight line can be interposed".* This is in fact, a somewhat loose way of phrasing the definition of the derivative we have given here. Why, then, did Fermat, Newton, and Leibniz change the emphasis from the method of exhaustion to the method of limits? The reason must lie in the computational power of limits, which enabled Newton and Leibniz to establish the rules of calculus, in spite of the fact that limits were not clearly understood for at least another century. However, there is nothing to prevent one from carrying out the same program using the method of exhaustion. We shall do so in this book.

In teaching by this approach, one may begin by defining transition points (with examples like birth, freezing, and sunset) and then go on to define overtaking and the derivative. (One must emphasize the fact that, in the definition of a transition point, nothing is said about the number x_0 itself.) The notion of transition point occurs again in graphing, when we consider turning points and inflection points, so the computational techniques needed to determine overtaking are put to good use later.

Since the definition of the derivative is so close to that of the integral (a transition point between lower and upper sums), the treatment of the fundamental theorem of calculus becomes very simple.

For those courses in which the completeness of the real numbers is emphasized, the following version of the completeness axiom is especially well suited to the transitions approach.

---

*C. Boyer, *The History of the Calculus and Its Conceptual Development*, Dover (1959), p. 57.

Definition A set A of real numbers is convex if, whenever x and y are in A and $x < z < y$, then z is in A.

Completeness Axiom Every convex set of real numbers is an interval.

With this axiom, the proofs of "hard" properties of continuous functions, such as their boundedness on closed intervals and their integrability, are within the reach of most first-year students.

Although the transition-point approach has some computational disadvantages, it does enable one to present a logically complete definition, with geometric and physical motivation at the end of only one hour of lecture. (Chapter 1 is such a lecture.) Coupled with an early intuitive approach to limits for their computational power, this method allows one to delay rigorous limits until later in the course when students are ready for them, and when they are really needed for topics like L'Hôpital's rule, improper integrals, and infinite senes.

Limits are so important in mathematics that they cannot be ignored in any calculus course. It is tempting to introduce them early because they are simple to use in calculations, but the subtlety of the limit concept often causes beginning students to feel uneasy about the foundations of calculus. Transitions, in contrast, provide conceptually simple definitions of the derivative and integral, but they are quite complicated to use in calculations. Fortunately, one does not need to do many calculations directly from the definition. The great "machine" of Newton and Leibniz enables us to calculate derivatives by a procedure which is independent of the particular form of the definition being used.

Although our reasons for using the method of transitions stem mostly from trying to make calculus easier to learn, we have another reason as well. Many mathematicians have complained that calculus gives a distorted picture of modern mathematics, with total emphasis on "analysis." We hope that the use of transitions will partly answer this complaint. It gives a better balance between the various disciplines of mathematics and gives the student a more accurate picture of what modern mathematics is all about.

We have mentioned that the concept of transition is important in its own right, and we gave some nonmathematical examples involving sudden changes. Although the notion of transition is built into differential and integral calculus, the classical techniques of calculus (limits, the rules of calculus, and so on) have proven insufficient as a tool for studying many discontinuous phenomena. Such phenomena are of common occurrence in biology and the social sciences and include, for example, revolutions, birth, and death. They may all be described as transitions.

Contemporary mathematicians have been paying more and more attention to discontinuous phenomena and a geometric or qualitative description of nature. In biology and sociology this aspect is an important complement to a quantitative analysis. Our emphasis on transitions is inspired partly by the belief that this concept will play an increasingly important role in the applications of mathematics.*

---

*Of particular interest in this direction is the catastrophe theory introduced by the French mathematician René Thom in his book *Structural Stability and Morphogenesis* (Benjamin, 1975). Thom's book is somewhat philosophical in nature, but more concrete applications of catastrophe theory can be found in the books: E. C. Zeeman, *Catastrophe Theory: Selected Papers 1972-1977* (Addison-Wesley, 1977); T. Poston and I. Stewart, *Catastrophe Theory and its Applications* (Fearon-Pitman, 1978). The applications to biological and social sciences have received some rather sharp criticism; the main critical paper, by H. J. Sussmann and R. S. Zahler, is "Catastrophe theory as applied to the social and biological sciences: a critique," *Synthese* **37** (1978), 117-216. A general account of the controversy is given in A. Woodcock and M. Davis, *Catastrophe Theory* (Dutton, 1978).

Contents

1 The Derivative

This chapter gives a complete definition of the derivative assuming a knowledge of high-school algebra, including inequalities, functions, and graphs. The next chapter will reformulate the definition in different language, and in Chapter 13 we will prove that it is equivalent to the usual definition in terms of limits.

The definition uses the concept of change of sign, so we begin with this.

Change of Sign

A function is said to *change sign* when its graph crosses from one side of the x axis to the other. We can define this concept precisely as follows.

Definition Let f be a function and x_0 a real number. We say that f *changes sign from negative to positive at* x_0 if there is an open interval (a, b) containing x_0 such that f is defined on (a, b) (except possibly at x_0) and

$$f(x) < 0 \qquad \text{if } a < x < x_0$$

and

$$f(x) > 0 \qquad \text{if } x_0 < x < b$$

Similarly, we say that *f changes sign from positive to negative at* x_0 if there is an open interval (a, b) containing x_0 such that f is defined on (a, b) (except possibly at x_0) and

$$f(x) > 0 \qquad \text{if } a < x < x_0$$

and

$$f(x) < 0 \qquad \text{if } x_0 < x < b$$

Notice that the interval (a, b) may have to be chosen small, since a function which changes sign from negative to positive may later change back from positive to negative (see Fig. 1-1).

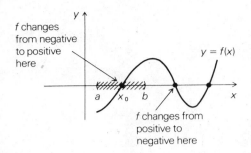

f changes
from negative
to positive
here

$y = f(x)$

f changes from
positive to
negative here

Fig. 1-1   Change of sign.

Worked Example 1   Where does $f(x) = x^2 - 5x + 6$ change sign?

Solution   We factor f and write $f(x) = (x - 3)(x - 2)$. The function f changes sign whenever one of its factors does. This occurs at $x = 2$ and $x = 3$. The factors have opposite signs for x between 2 and 3, and the same sign otherwise, so f changes from positive to negative at $x = 2$ and from negative to positive at $x = 3$. (See Fig. 1-2).

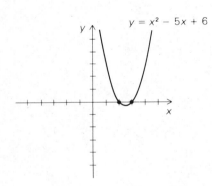

$y = x^2 - 5x + 6$

Fig. 1-2 $y = x^2 - 5x + 6$
changes sign at $x = 2$ and
$x = 3$.

We can compare two functions, f and g, by looking at the sign changes of the difference $f(x) - g(x)$. The following example illustrates the idea.

Worked Example 2   Let $f(x) = \frac{1}{2}x^3 - 1$ and $g(x) = x^2 - 1$.
(a) Find the sign changes of $f(x) - g(x)$.
(b) Sketch the graphs of f and g on the same set of axes.
(c) Discuss the relation between the results of parts (a) and (b).

Solution
(a) $f(x) - g(x) = \frac{1}{2}x^3 - 1 - (x^2 - 1) = \frac{1}{2}x^3 - x^2 = \frac{1}{2}x^2(x - 2)$. Since the factor x appears twice, there is no change of sign at $x = 0$ (x^2 is positive both for $x < 0$ and for $x > 0$). There is a change of sign from negative to positive at $x = 2$.

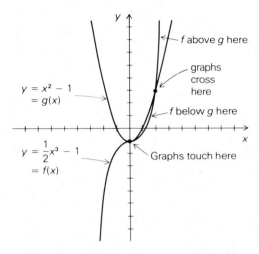

$y = x^2 - 1$
$= g(x)$

$y = \dfrac{1}{2}x^3 - 1$
$= f(x)$

f above g here

graphs
cross
here

f below g here

Graphs touch here

Fig. 1-3 $f - g$ changes sign when the graphs of f and g cross.

(b) See Fig. 1-3.

(c) Since $f(x) - g(x)$ changes sign from negative to positive at $x = 2$, we can say:

$$\text{If } x \text{ is near 2 and } x < 2, \text{ then } f(x) - g(x) < 0; \text{ that is, } f(x) < g(x).$$

and

$$\text{If } x \text{ is near 2 and } x > 2, \text{ then } f(x) - g(x) > 0; \text{ that is, } f(x) > g(x).$$

Thus the graph of f must cross the graph of g at $x = 2$, passing from below to above it as x passes 2.

Solved Exercises*

1. If $f(x)$ is a polynomial and $f(x_0) = 0$, must f necessarily change sign at x_0?

2. For which positive integers n does $f(x) = x^n$ change sign at zero?

3. If $r_1 \neq r_2$, describe the sign change at r_1 of $f(x) = (x - r_1)(x - r_2)$.

Exercises

1. Find the sign changes of each of the following functions:

(a) $f(x) = 2x - 1$ (b) $f(x) = x^2 - 1$

(c) $f(x) = x^2$ (d) $h(z) = z(z - 1)(z - 2)$

*Solutions appear in the Appendix.

2. Describe the change of sign at $x = 0$ of the function $f(x) = mx$ for $m = -2$, $0, 2$.

3. Describe the change of sign at $x = 0$ of the function $f(x) = mx - x^2$ for $m = -1, -\frac{1}{2}, 0, \frac{1}{2}, 1$.

4. Let $f(t)$ denote the angle of the sun above the horizon at time t. When does $f(t)$ change sign?

Estimating Velocities

If the position of an object moving along a line changes linearly with time, the object is said to be in *uniform motion*, and the rate of change of position with respect to time is called the *velocity*. The velocity of a uniformly moving object is a fixed number, independent of time. Most of the motion we observe in nature is not uniform, but we still feel that there is a quantity which expresses the rate of movement at any instant of time. This quantity, which we may call the *instantaneous velocity*, will depend on the time.

To illustrate how instantaneous velocity might be estimated, let us suppose that we are looking down upon a car C which is moving along the middle lane of a three-lane, one-way road. Without assuming that the motion of the car is uniform, we wish to estimate the velocity v_0 of the car at exactly 3 o'clock.

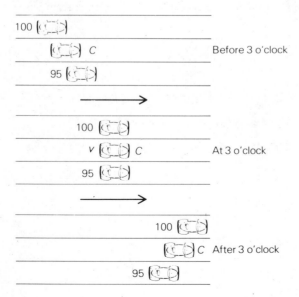

Fig. 1-4 The velocity of car C is between 95 and 100 kilometers per hour.

Suppose that we have the following information (see Fig. 1-4): A car which was moving uniformly with velocity 95 kilometers per hour was passed by car C at 3 o'clock, and a car which was moving uniformly with velocity 100 kilometers per hour passed car C at 3 o'clock.

We conclude that v_0 was at least 95 kilometers per hour and at most 100 kilometers per hour. This estimate of the velocity could be improved if we were to compare car x with more "test cars."

In general, let the variable y represent distance along a road (measured in kilometers from some reference point) and let x represent time (in hours from some reference moment). Suppose that the position of two cars traveling in the positive direction is represented by functions $f_1(x)$ and $f_2(x)$. Then car 1 passes car 2 at time x_0 if the function $f_1(x) - f_2(x)$, which represents the "lead" of car 1 over car 2, changes sign from negative to positive at x_0. (See Fig. 1-5.) When this happens, we expect car 1 to have a higher velocity than car 2 at time x_0.

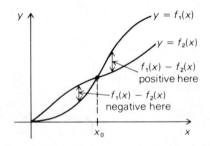

Fig. 1-5 $f_1 - f_2$ changes sign from negative to positive at x_0.

Since the graph representing uniform motion with velocity v is a straight line with slope v, we could estimate the velocity of a car whose motion is nonuniform by seeing how the graph of the function giving its position crosses straight lines with various slopes.

Worked Example 3 Suppose that a moving object is at position $y = f(x) = \frac{1}{2}x^2$ at time x. Show that its velocity at $x_0 = 1$ is at least $\frac{1}{2}$.

Solution We use a "test object" whose velocity is $v = \frac{1}{2}$ and whose position at time x is $\frac{1}{2}x$. When $x = x_0 = 1$, both objects are at $y = \frac{1}{2}$. When $0 < x < 1$, we have $x^2 < x$, so $\frac{1}{2}x^2 < \frac{1}{2}x$; when $x > 1$, we have $\frac{1}{2}x^2 > \frac{1}{2}x$. It follows that the difference $\frac{1}{2}x^2 - \frac{1}{2}x$ changes sign from negative to positive at 1, so the velocity of our moving object is at least $\frac{1}{2}$ (see Fig. 1-6).

Solved Exercise

---

4. Show that the velocity at $x_0 = 1$ of the object in Worked Example 3 is at most 2.

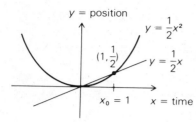

Fig. 1-6 The graph of $y = \frac{1}{2}x$ is above that of $y = \frac{1}{2}x^2$ when $0 < x < 1$ and is below when $x > 1$.

Exercise

5. How does the velocity at $x_0 = 1$ of the object in Worked Example 3 compare with $\frac{3}{4}$? With $\frac{3}{2}$?

Definition of the Derivative

While keeping the idea of motion and velocity in mind, we will continue our discussion simply in terms of functions and their graphs. Recall that the line through (x_0, y_0) with slope m has the equation $y - y_0 = m(x - x_0)$. Solving for y in terms of x, we find that this line is the graph of the linear function $l(x) = y_0 + m(x - x_0)$. We can estimate the "slope" of a given function $f(x)$ at x_0 by comparing $f(x)$ and $l(x)$, i.e. by looking at the sign changes at x_0 of $f(x) - l(x) = f(x) - [f(x_0) + m(x - x_0)]$ for various values of m. Here is a precise formulation.

Definition Let f be a function whose domain contains an open interval about x_0. We say that the number m_0 is the *derivative of f at x_0*, provided that:

1. For every $m < m_0$, the function

$$f(x) - [f(x_0) + m(x - x_0)]$$

changes sign from negative to positive at x_0.

2. For every $m > m_0$, the function

$$f(x) - [f(x_0) + m(x - x_0)]$$

changes sign from positive to negative at x_0.

If such a number m_0 exists, we say that f is *differentiable at* x_0, and we write $m_0 = f'(x_0)$. If f is differentiable at each point of its domain, we just say that f is *differentiable*. The process of finding the derivative of a function is called *differentiation*.

Geometrically, the definition says that lines through $(x_0, f(x_0))$ with slope less than $f'(x_0)$ cross the graph of f from above to below, while lines with slope greater than $f'(x_0)$ cross from below to above. (See Fig. 1-7.)

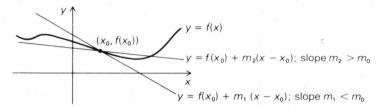

Fig. 1-7 Lines with slope different from m_0 cross the curve.

Given f and x_0, the number $f'(x_0)$ is uniquely determined if it exists. That is, at most one number satisfies the definition. Suppose that m_0 and $\overline{m}_0$ both satisfied the definition, and $m_0 \neq \overline{m}_0$; say $m_0 < \overline{m}_0$. Let $m = (m_0 + \overline{m}_0)/2$, so $m_0 < m < \overline{m}_0$. By condition 1 for $\overline{m}_0$, $f(x) - [f(x_0) + m(x - x_0)]$ changes sign from negative to positive at x_0, and by condition 2 for m_0, it changes sign from positive to negative at x_0. But it can't do both! Therefore $m_0 = \overline{m}_0$.

The line through $(x_0, f(x_0))$, whose slope is exactly $f'(x_0)$ is pinched, together with the graph of f, between the "downcrossing" lines with slope less than $f'(x_0)$ and the "upcrossing" lines with slope greater than $f'(x_0)$. It is the line with slope $f'(x_0)$, then, which must be tangent to the graph of f at (x_0, y_0). We may take this as our *definition* of tangency. (See Fig. 1-8.)

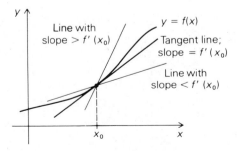

Fig. 1-8 The slope of the tangent line is the derivative.

Definition Suppose that the function f is differentiable at x_0. The line $y = f(x_0) + f'(x_0)(x - x_0)$ through $(x_0, f(x_0))$ with slope $f'(x_0)$ is called the *tangent line to the graph of f at* $(x_0, f(x_0))$.

Following this definition, we sometimes refer to $f'(x_0)$ as the *slope of the curve* $y = f(x)$ at the point $(x_0, f(x_0))$. Note that the definitions do not say anything about how (or even whether) the tangent line itself crosses the graph of a function. (See Problem 7 at the end of this chapter.)

Recalling the discussion in which we estimated the velocity of a car by seeing which cars it passed, we can now give a mathematical definition of velocity.

Definition Let $y = f(x)$ represent the position at time x of a moving object. If f is differentiable at x_0, the number $f'(x_0)$ is called the (instantaneous) *velocity* of the object at the time x_0.

More generally, we call $f'(x_0)$ the *rate of change* of y with respect to x at x_0.

Worked Example 4 Find the derivative of $f(x) = x^2$ at $x_0 = 3$. What is the equation of the tangent line to the parabola $y = x^2$ at the point $(3,9)$?

Solution According to the definition of the derivative—with $f(x) = x^2, x_0 = 3$, and $f(x_0) = 9$—we must investigate the sign change at 3, for various values of m, of

$$
\begin{aligned}
f(x) - [f(x_0) + m(x - x_0)] &= x^2 - [9 + m(x - 3)] \\
&= x^2 - 9 - m(x - 3) \\
&= (x + 3)(x - 3) - m(x - 3) \\
&= (x - 3)(x + 3 - m)
\end{aligned}
$$

According to Solved Exercise 3, with $r_1 = 3$ and $r_2 = m - 3$, the sign change is:

1. From negative to positive if $m - 3 < 3$; that is, $m < 6$.

2. From positive to negative if $3 < m - 3$; that is, $m > 6$.

We see that the number $m_0 = 6$ fits the conditions in the definition of the derivative, so $f'(3) = 6$. The equation of the tangent line at $(3,9)$ is therefore $y = 9 + 6(x - 3)$; that is, $y = 6x - 9$. (See Fig. 1-9.)

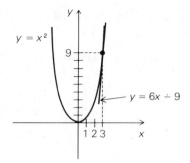

Fig. 1-9 The equation of
the line tangent to $y = x^2$
at $x_0 = 3$ is $y = 6x - 9$.

Solved Exercises

5. Let $f(x) = x^3$. What is $f'(0)$? What is the tangent line at $(0,0)$?

6. Let f be a function for which we know that $f(3) = 2$ and $f'(3) = \sqrt[5]{8}$. Find the y intercept of the line which is tangent to the graph of f at $(3,2)$.

7. Let $f(x) = |x| = \begin{cases} x & \text{if } x \geqslant 0 \\ -x & \text{if } x < 0 \end{cases}$ (the absolute value function). Show by a geometric argument that f is not differentiable at zero.

8. The position of a moving object at time x is x^2. What is the velocity of the object when $x = 3$?

Exercises

6. Find the derivative of $f(x) = x^2$ at $x = 4$. What is the equation of the tangent line to the parabola $y = x^2$ at $(4, 16)$?

7. If $f(x) = x^4$, what is $f'(0)$?

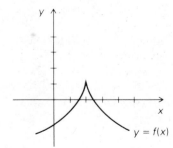

Fig. 1-10 Where is f differentiable? (See Exercise 9.)

8. The position at time x of a moving object is x^3. What is the velocity at $x = 0$?

9. For which value of x_0 does the function in Fig. 1-10 fail to be differentiable?

The Derivative as a Function

The preceding examples show how derivatives may be calculated directly from the definition. Usually, we will not use this cumbersome method; instead, we will use *differentiation rules*. These rules, once derived, enable us to differentiate many functions quite simply. In this section, we will content ourselves with deriving the rules for differentiating linear and quadratic functions. General rules will be introduced in Chapter 3.

The following theorem will enable us to find the tangent line to any parabola at any point.

Theorem 1   Quadratic Function Rule. *Let* $f(x) = ax^2 + bx + c$, *where* a, b, *and* c *are constants, and let* x_0 *be any real number. Then* f *is differentiable at* x_0, *and* $f'(x_0) = 2ax_0 + b$.

Proof We must investigate the sign changes at x_0 of the function

$$f(x) - [f(x_0) + m(x - x_0)]$$
$$= ax^2 + bx + c - [ax_0^2 + bx_0 + c + m(x - x_0)]$$
$$= a(x^2 - x_0^2) + b(x - x_0) - m(x - x_0)$$
$$= (x - x_0)[a(x + x_0) + b - m]$$

The factor $[a(x + x_0) + b - m]$ is a (possibly constant) linear function whose value at $x = x_0$ is $a(x_0 + x_0) + b - m = 2ax_0 + b - m$. If $m < 2ax_0 + b$, this factor is positive at $x = x_0$, and being a linear function it is also positive when x is near x_0. Thus the product of $[a(x + x_0) + b - m]$ with $(x - x_0)$ changes sign from negative to positive at x_0. If $m > 2ax_0 + b$, then the factor $[a(x + x_0) + b - m]$ is negative when x is near x_0, so its product with $(x - x_0)$ changes sign from positive to negative at x_0.

Thus the number $m_0 = 2ax_0 + b$ satisfies the definition of the derivative, and so $f'(x_0) = 2ax_0 + b$.

Worked Example 5 Find the derivative at -2 of $f(x) = 3x^2 + 2x - 1$.

Solution Applying the quadratic function rule with $a = 3$, $b = 2$, $c = -1$, and $x_0 = -2$, we find $f'(-2) = 2(3)(-2) + 2 = -10$.

We can use the quadratic function rule to obtain quickly a fact which may be known to you from analytic geometry.

Worked Example 6 Suppose that $a \neq 0$. At which point does the parabola $y = ax^2 + bx + c$ have a horizontal tangent line?

Solution The slope of the tangent line through the point $(x_0, ax_0^2 + bx_0 + c)$ is $2ax_0 + b$. This line is horizontal when its slope is zero; that is, when $2ax_0 + b = 0$, or $x_0 = -b/2a$. The y value here is $a(-b/2a)^2 + b(-b/2a) + c = b^2/4a - b^2/2a + c = -(b^2/4a) + c$. The point $(-b/2a, -b^2/4a + c)$ is called the *vertex* of the parabola $y = ax^2 + bx + c$.

In Theorem 1 we did not require that $a \neq 0$. When $a = 0$, the function $f(x) = ax^2 + bx + c$ is linear, so we have the following corollary:

Corollary Linear Function Rule. *If $f(x) = bx + c$, and x_0 is any real number, then $f'(x_0) = b$.*

In particular, if $f(x) = c$, a constant function, then $f'(x_0) = 0$ for all x_0.

For instance, if $f(x) = 3x + 4$, then $f'(x_0) = 3$ for any x_0; if $g(x) = 4$, then $g'(x_0) = 0$ for any x_0.

This corollary tells us that the rate of change of a linear function is just the slope of its graph. Note that it does not depend on x_0: the rate of change of a linear function is *constant*. For a general quadratic function, though, the derivative $f'(x_0)$ does depend upon the point x_0 at which the derivative is taken. In fact, we can consider f' as a *new function*; writing the letter x instead of x_0, we have $f'(x) = 2ax + b$.

Definition Let f be any function. We define the function f', with domain equal to the set of points at which f is differentiable, by setting $f'(x)$ equal to the derivative of f at x. The function $f'(x)$ is simply called the *derivative* of $f(x)$.

Worked Example 7 What is the derivative of $f(x) = 3x^2 - 2x + 1$?

Solution By the quadratic function rule, $f'(x_0) = 2 \cdot 3x_0 - 2 = 6x_0 - 2$.

Writing x instead of x_0, we find that the derivative of $f(x) = 3x^2 - 2x + 1$ is $f'(x) = 6x - 2$.

Remark When we are dealing with functions given by specific formulas, we often omit the function names. For example, we could state the result of Worked Example 7 as "the derivative of $3x^2 - 2x + 1$ is $6x - 2$."

Since the derivative of a function f is another function f', we can go on to differentiate f' again. The result is yet another function, called the *second derivative* of f and denoted by f''.

Worked Example 8 Find the second derivative of $f(x) = 3x^2 - 2x + 1$.

Solution We must differentiate $f'(x) = 6x - 2$. This is a linear function; applying the formula for the derivative of a linear function, we find $f''(x) = 6$. The second derivative of $3x^2 - 2x + 1$ is thus the constant function whose value for every x is equal to 6.

If $f(x)$ is the position of a moving object at time x, then $f'(x)$ is the velocity, so $f''(x)$ is the rate of change of velocity with respect to time. It is called the *acceleration* of the object.

We end with a remark on notation. It is not necessary to represent functions by f and independent and dependent variables by x and y; as long as we say what we are doing, we can use any letters we wish.

Worked Example 9 Let $g(a) = 4a^2 + 3a - 2$. What is $g'(a)$? What is $g'(2)$?

Solution If $f(x) = 4x^2 + 3x - 2$, we know that $f'(x) = 8x + 3$. Using g instead of f and a instead of x, we have $g'(a) = 8a + 3$. Finally, $g'(2) = 8 \cdot 2 + 3 = 19$.

Solved Exercises

---

9. Let $f(x) = 3x + 1$. What is $f'(8)$?

10. An apple falls from a tall tree toward the earth. After t seconds, it has fallen $4.9t^2$ meters. What is the velocity of the apple when $t = 3$? What is the acceleration?

11. Find the equation of the line tangent to the graph of $f(x) = 3x^2 + 4x + 2$ at the point where $x_0 = 1$.

12. For which functions $f(x) = ax^2 + bx + c$ is the second derivative equal to the zero function?

Exercises

10. Differentiate the following functions:

   (a) $f(x) = x^2 + 3x - 1$ (b) $f(x) = (x - 1)(x + 1)$
   (c) $f(x) = -3x + 4$ (d) $g(t) = -4t^2 + 3t + 6$

11. A ball is thrown upward at $t = 0$; its height in meters until it strikes the ground is $24.5t - 4.9t^2$ where the time is t seconds. Find:

   (a) The velocity at $t = 0, 1, 2, 3, 4, 5$.
   (b) The time when the ball is at its highest point.
   (c) The time when the velocity is zero.
   (d) The time when the ball strikes the ground.

12. Find the tangent line to the parabola $y = x^2 - 3x + 1$ when $x_0 = 2$. Sketch.

13. Find the second derivative of each of the following:

   (a) $f(x) = x^2 - 5$ (b) $f(x) = x - 2$
   (c) A function whose derivative is $3x^2 - 7$.

Problems for Chapter 1

1. Find the sign changes of:
   (a) $f(x) = (3x^2 - 1)/(3x^2 + 1)$ (b) $f(x) = 1/x$

2. Where do the following functions change sign from positive to negative?
   (a) $f(x) = 6 - 5x$ (b) $f(x) = 2x^2 - 4x + 2$
   (c) $f(x) = 2x - x^2$ (d) $f(x) = 6x + 1$
   (e) $f(x) = (x - 1)(x + 2)^2(x + 3)$

3. The position at time x of a moving object is x^3. Show that the velocity at time 1 lies between 2 and 4.

4. Let $f(x) = (x - r_1)^{n_1}(x - r_2)^{n_2} \cdots (x - r_k)^{n_k}$, where $r_1 < r_2 < \cdots < r_k$ are the roots of f and $n_1, \ldots, n_k$ are positive integers. Where does $f(x)$ change sign from negative to positive?

5. Using the definition of the derivative directly, find $f'(2)$ if $f(x) = 3x^2$.

6. If $f(x) = x^5 + x$, is $f'(0)$ positive or negative? Why?

7. Sketch each of the following graphs together with its tangent line at $(0, 0)$:
   (a) $y = x^2$ (b) $y = x^3$ (c) $y = -x^3$. Must a tangent line to a graph always lie on one side of the graph?

8. Find the derivative at $x_0 = 0$ of $f(x) = x^3 + x$.

9. Find the following derivatives:
   (a) $f(x) = x^2 - 2$; find $f'(3)$.
   (b) $f(x) = 1$; find $f'(7)$.
   (c) $f(x) = -13x^2 - 9x + 5$; find $f'(1)$.
   (d) $g(s) = 0$; find $g'(3)$.
   (e) $k(y) = (y + 4)(y - 7)$; find $k'(-1)$.
   (f) $x(f) = 1 - f^2$; find $x'(0)$.
   (g) $f(x) = -x + 2$; find $f'(3.752764)$.

10. Find the tangent line to the curve $y = x^2 - 2x + 1$ when $x = 2$. Sketch.

11. Let $f(x) = 2x^2 - 5x + 2$, $k(x) = 3x - 4$, $g(x) = \frac{3}{4}x^2 + 2x$, $l(x) = -2x + 3$, and $h(x) = -3x^2 + x + 3$.
    (a) Find the derivative of $f(x) + g(x)$ at $x = 1$.
    (b) Find the derivative of $3f(x) - 2h(x)$ at $x = 0$.
    (c) Find the equation of the tangent line to the graph of $f(x)$ at $x = 1$.
    (d) Where does $l(x)$ change sign from negative to positive?
    (e) Where does $l(x) - [k(x) - k'(-1)](x + 1)$ change sign from positive to negative?

12. Find the tangent line to the curve $y = -3x^2 + 2x + 1$ when $x = 0$. Sketch.

13. Let R be any point on the parabola $y = x^2$. (a) Draw the horizontal line through R. (b) Draw the perpendicular to the tangent line at R. Show that the distance between the points where these lines cross the y axis is equal to $\frac{1}{2}$, regardless of the value of x. (Assume, however, that $x \neq 0$.)

14. Given a point $(\bar{x}, \bar{y})$, find a general rule for determining how many lines through the point are tangent to the parabola $y = x^2$.

15. If $f(x) = ax^2 + bx + c = a(x - r)(x - s)$ (r and s are the roots of f), show that the values of $f'(x)$ at r and s are negatives of one another. Explain this by appeal to the symmetry of the graph.

16. Let $f(t) = 2t^2 - 5t + 2$ be the position of object A and let $h(t) = -3t^2 + t + 3$ be the position of object B.
    (a) When is A moving faster than B?
    (b) How fast is B going when A stops?
    (c) When does B change direction?

17. Let $f(x) = 2x^2 + 3x + 1$.
    (a) For which values of x is $f'(x)$ negative, positive, and zero?
    (b) Identify these points on a graph of f.

18. How do the graphs of functions $ax^2 + bx + c$ whose second derivative is positive compare with those for which the second derivative is negative and those for which the second derivative is zero?

19. Where does the function $f(x) = -2|x - 1|$ fail to be differentiable? Explain your answer with a sketch.

2 Transitions and Derivatives

In this chapter we reformulate the definition of the derivative in terms of the concept of transition point. Other concepts, such as change of sign, can also be expressed in terms of transitions. In addition to the new language, a few new basic properties of change of sign and overtaking will be introduced.

In what follows the reader is assumed to be familiar with the interval notation and the containment symbol $\in$. Thus $x \in (a, b]$ means $a < x \leqslant b$, $x \in (-\infty, b)$ means $x < b$, etc. Intervals of the form (a, b) are called open, while those of the form $[a, b]$ are called closed.

Transition Points

Some changes are sudden, or definitive, and are marked by a transition point. The time of sunrise marks the transition from night to day, and the summer solstice marks the transition from spring to summer. Not all transitions take place in time, though. For example, let T denote the temperature of some water. For certain values of T, the water is in a liquid state; for other values of T, the water is in a solid state (ice) or a gaseous state (water vapor). Between these states are two transition temperatures, the freezing point and the boiling point.

Here is another example. A tortoise and a hare are running a race. Let T denote the period of time during which the tortoise is in the lead; let H denote the period of time during which the hare is ahead. A moment at which the hare overtakes the tortoise is a transition point from T to H. When the tortoise overtakes the hare, the transition is from H to T.

In order to do mathematics with the concept of transition, we must give a formal definition. The following definition has been chosen for its intuitive content and for technical convenience.

Definition Let A and B be two sets of real numbers. A number x_0 is called a *transition point from A to B* if there is an open interval I containing x_0 such that

1. If $x \in I$ and $x < x_0$, then x is in A but not in B.

2. If $x \in I$ and $x > x_0$, then x is in B but not in A.

There are several remarks to be made concerning this definition. The first one concerns definitions in general; the others concern the particular definition above.

Remark 1 Definitions play an important role in mathematics. They set out in undisputable terms what is meant by a certain phrase, such as "x_0 is a transition point from A to B." Definitions are usually made to reflect some intuitive idea, and our intuition is usually a reliable guide to the use of the defined expression. Still, if we wish to establish that "x_0 is a point of transition from A to B" in a given example, the definition is the final authority; we must demonstrate that the conditions set out in the definition are met. Partial or approximate compliance is not acceptable; the conditions must be met fully and exactly.

There is no disputing the correctness of a definition, but only its usefulness. Over a long period of mathematical history, the most useful definitions have survived. Thousands have been discarded as inappropriate or useless.

Remark 2 There are a couple of specific points to be noted in the definition above. First of all, we do not specify to what set the point x_0 itself belongs. It may belong to A, B, both, or neither. The second thing to notice is the role of the interval I. Its inclusion in the definition corresponds to the intuitive notion that transitional change may be temporary. For instance, in the transition from ice to water, the interval I must be chosen so that its right-hand endpoint is less than or equal to the boiling point. Reread the definition now to be sure that you understand this remark.

Remark 3 In the example of the tortoise and the hare, suppose that the tortoise is behind for $t < t_1$, that they run neck and neck from t_1 to t_2, and that the tortoise leads for $t > t_2$. There is no transition *point* in this case, but rather a transition *period*. In our definition, we are only concerned with transition points. In nonmathematical situations, it is not always clear when a transition is abrupt and when it occurs over a period. Consider, for example, the transition of power from one government to another, or the transition of an embryo from pre-life to life.

Worked Example 1 Let A be the set of real numbers r for which the point $(1,2)$ lies *outside* the circle $x^2 + y^2 = r^2$. Let B be the set of r for which $(1,2)$ lies inside $x^2 + y^2 = r^2$. Find the transition point from A to B. Does it belong to A, B, both, or neither?

Solution The point $(1,2)$ lies at a distance $\sqrt{1^2 + 2^2} = \sqrt{5}$ from the origin, so $A = (0,\sqrt{5})$ and $B = (\sqrt{5},\infty)$. The transition point is $\sqrt{5}$, which belongs to neither A nor B. (The point $(1, 2)$ lies *on* the circle of radius $\sqrt{5}$, not inside or outside it.)

If A and B are intervals, then a transition point from A to B must be a common endpoint. (See Fig. 2-1.) Notice that, when A and B are intervals, there is at most one transition point between them, which may or may not belong to A or B.

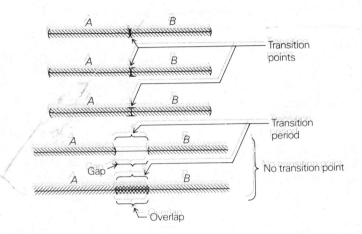

Fig. 2-1 If A and B are intervals, transition points are common endpoints.

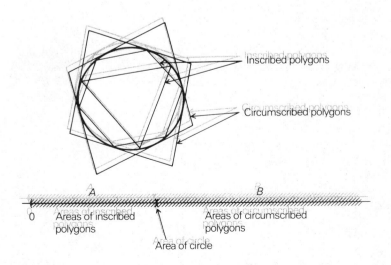

Fig. 2-2 The area of a circle can be described as a transition point.

Some transitions occur along curves or surfaces. For example, the coast-line is the transition curve between land and sea, while your skin is the transition surface between your body and the atmosphere. We give a mathematical example of a transition curve in Chapter 6, but we will usually be dealing with transition points.

The notion of transition occurs in the answer given by the ancient Greeks to the question, "What is the area enclosed by a circle of radius r?" We may consider the set A consisting of the areas of all possible polygons inscribed in the circle and the set B consisting of the areas of all possible circumscribed polygons. The transition point from A to B is the area of the circle. (See Fig. 2-2.) By using inscribed and circumscribed regular polygons with sufficiently many sides, Archimedes was able to locate the transition point quite accurately.

Solved Exercises*

1. Let B be the set consisting of those x for which $x^2 - 1 > 0$, and let $A = (-1, 1]$. Find the transition points from B to A and from A to B.

2. Let $A = (-\infty, -1/1000)$, $B = (1/1000, \infty)$, $C = [-1/1000, 1/1000]$. Find the transition points from each of these sets to each of the others.

3. Let x be the distance from San Francisco on a road crossing the United States. If A consists of those x for which the road is in California at mile x, and B consists of those x for which the road is in Nevada at mile x, what is the transition point called?

Exercises

1. Describe the following as transition points:

    (a) Vernal equinox.   (b) Entering a house.

    (c) A mountain top.   (d) Zero.

    (e) Outbreak of war.   (f) Critical mass.

    (g) A window shattering.   (h) Revolution.

2. What transition points can you identify in the following phenomena? Describe them.

    (a) The movement of a pendulum.   (b) Diving into water.

    (c) A traffic accident.   (d) Closing a door.

    (e) Drinking a glass of water.   (f) Riding a bicycle.

*Solutions appear in the Appendix.

3. For each of the following pairs of functions, $f(x)$ and $g(x)$, let A be the set of x where $f(x) > g(x)$, and let B be the set of x where $f(x) < g(x)$. Find the transition points, if there are any, from A to B and from B to A.

   (a) $f(x) = 2x - 1$; $g(x) = -x + 2$ (b) $f(x) = x^2 + 2$; $g(x) = 3x + 6$

   (c) $f(x) = x^3 - x$; $g(x) = x$ (d) $f(x) = x^2 - 1$; $g(x) = -x^2 + 1$

4. Find the transition points from A to B and from B to A in each of the following cases:

   (a) $A = [1,3)$

   $B =$ the set of x for which $x^2 - 4x + 3 > 0$

   (b) $A =$ the set of x for which $-3 < x \leqslant 1$ or $10 < x \leqslant 15$

   $B =$ the set of x for which $x < -3, 0 \leqslant x \leqslant 10$, or $16 < x$

Change of Sign and Overtaking

The concept of change of sign was defined in Chapter 1. Now we express it in terms of transitions. Let f be any function, N the set of x for which x is in the domain of f and $f(x) < 0$, and P the set of x for which x is in the domain of f and $f(x) > 0$.

Theorem 1 *(a) x_0 is a point of transition from N to P if and only if f changes sign from negative to positive at x_0. (b) x_0 is a point of transition from P to N if and only if f changes sign from positive to negative at x_0.*

Proof (a) Suppose that x_0 is a transition point from N to P. By definition, there is an open interval I containing x_0 such that (i) if $x \in I$ and $x < x_0$, then x is in N but not in P, and (ii) if $x \in I$ and $x > x_0$, then x is in P but not in N. Letting $I = (a, b)$ and noting that $x \in (a, b)$ and $x < x_0$ is the same as saying $a < x < x_0$, we see that (i) reads: if $a < x < x_0$ then x is in the domain of f and $f(x) < 0$; and (ii) reads: if $x_0 < x < b$ then x is in the domain of f and $f(x) > 0$. Thus, by the definition of change of sign given in Chapter 1, f changes sign from negative to positive. Conversely, we can reverse this argument to show that if f changes sign from negative to positive, then x_0 is a transition point from P to N.

The proof of (b) is similar.

Let us return to the race between the tortoise and the hare. Denote by $f(t)$ the tortoise's position at time t and by $g(t)$ the hare's position at time t. The transition *"the tortoise overtakes the hare at time t_0"* means that there is an open interval I about t_0 such that:

1. If $t \in I$ and $t < t_0$, then $f(t) < g(t)$.

2. If $t \in I$ and $t > t_0$, then $f(t) > g(t)$.

If we graph f and g, this means that the graph of f lies below that of g for t just to the left of t_0 and above that of g for t just to the right of t_0. (See Fig. 2-3.) Notice the role of the interval I. (We could have taken a slightly larger one.) It appears in the definition because the hare may overtake the tortoise at a later time t_1.

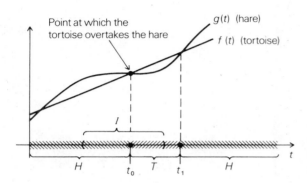

Point at which the tortoise overtakes the hare

$g(t)$ (hare)

$f(t)$ (tortoise)

Fig. 2-3 The tortoise overtakes the hare at t_0 and the hare overtakes the tortoise at t_1.

We now state the definition of overtaking for general functions.

Definition Let f and g be two functions, A the set of x (in the domain of f and of g) such that $f(x) < g(x)$, and B the set of x (in the domain of f and g) such that $f(x) > g(x)$. If x_0 is a transition point from A to B, we say that *f overtakes g* at x_0.

In other words, f overtakes g at x_0 if there is an interval I containing x_0 such that (f and g are defined on I, except possibly at x_0) and

1. For x in I and $x < x_0, f(x) < g(x)$.

2. For x in I and $x > x_0, f(x) > g(x)$.

We call an open interval I about x_0 for which conditions 1 and 2 are true *an interval which works* for f and g at x_0; i.e., I is small enough so that in I to the left of x_0, f is below g, while in I to the right of x_0, f is above g. (See Fig. 2-4.) Clearly, if a certain interval I works for f and g at x_0, so does any open interval J contained in I, as long as it still contains x_0.

If the tortoise and hare both fall asleep and start running the wrong way when they wake up, we still say that "the tortoise overtakes the hare" if the hare passes the tortoise when going in the wrong direction. (See Fig. 2-5.) In the general situation, when f overtakes g at x_0, the graph of f may actually be going downward. It is only the change in f as compared with the change in g which is important.

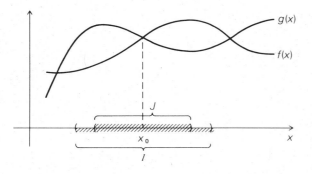

Fig. 2-4 If I is an interval that works (for f overtaking g at x_0), so is any smaller interval J.

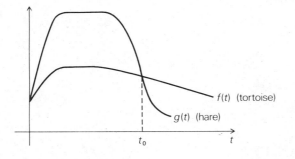

Fig. 2-5 The tortoise overtakes the hare (f overtakes g) at t_0.

Worked Example 2   For the functions f and g in Fig. 2-6, tell whether f overtakes g, g overtakes f; or neither, at each of the points x_1, x_2, x_3, x_4, and x_5. When overtaking takes place, indicate an interval which works.

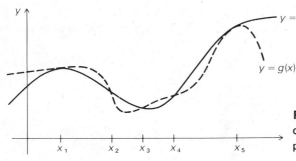

Fig. 2-6 What overtakings occur at the indicated points?

Solution   Neither function overtakes the other at x_1. In fact, in the interval $(0, x_2)$ about x_1, $g(x) > f(x)$ for x both to the right and to the left of x_1. At x_2, f overtakes g; an interval which works is $I = (x_1, x_3)$. (Note that the number x_2 is what we called x_0 in the definition.) At x_3, g overtakes f; an interval which works is (x_2, x_4). At x_4, f overtakes g again; an interval which works is (x_3, x_5). Neither function overtakes the other at x_5.

The relations between the concepts of overtaking and change of sign are explored in Problems 8 and 9 at the end of the chapter.

If, while the tortoise is overtaking the hare at t_0, a snail overtakes the tortoise at t_0, then we may conclude that the snail overtakes the hare at t_0. Let us state this as a theorem about functions.

Theorem 2 *Suppose f, g, and h are functions such that f overtakes g at x_0 and g overtakes h at x_0. Then f overtakes h at x_0.*

Proof Let I_1 be an interval which works for f and g at x_0, and let I_2 be an interval which works for g and h at x_0. That these intervals exist follows from the assumptions of the theorem. Choose I to be any open interval about x_0 which is contained in both I_1 and I_2. For instance, you could choose I to consist of all points which belong to both I_1 and I_2. (Study Fig. 2-7, where a somewhat smaller interval is chosen. Although the three graphs intersect in a complicated way, notice that the picture looks quite simple in the shaded region above the interval I. You should return to the figure frequently as you read the rest of this proof.)

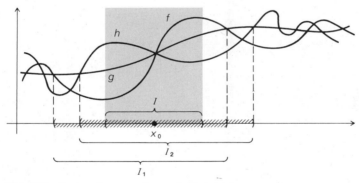

Fig. 2-7 *f* overtakes *g* and *g* overtakes *h* at x_0.

We will show that the interval I works for f and h. We begin by assuming that

$$x \in I \quad \text{and} \quad x < x_0 \tag{A}$$

Since I is contained in both I_1 and I_2, (A) implies:

$$x \in I_1 \quad \text{and} \quad x < x_0 \tag{A$_1$}$$
$$x \in I_2 \quad \text{and} \quad x < x_0 \tag{A$_2$}$$

Since I_1 works for f and g at x_0 and I_2 works for g and h at x_0, (A_1) and (A_2) imply

$f(x) < g(x)$ (and x is in the domain of f and g) (B_1)

$g(x) < h(x)$ (and x is in the domain of g and h) (B_2)

(B_1) and (B_2) together imply that

$f(x) < h(x)$ (and x is in the domain of f and h) (B)

This chain of reasoning began with (A) and concluded with (B), so we have proven that if $x \in I$ and $x < x_0$, then $f(x) < h(x)$. Similarly, one proves that if $x \in I$ and $x > x_0$ then $f(x) > h(x)$, so I works.

The next result provides a link between the concepts of linear change and transition. We know that a faster object overtakes a slower one when they meet. Here is the formal version of that fact for uniform motion. Its proof is a good exercise in the algebra of inequalities.

Theorem 3 *Let $f_1(x) = m_1x + b_1$ and $f_2(x) = m_2x + b_2$ be linear functions whose graphs both pass through the point (x_0, y_0). If $m_2 > m_1$, then f_2 overtakes f_1 at x_0.*

Proof Since $y_0 = f_1(x_0) = m_1x_0 + b_1$, we can solve for b_1 to get $b_1 = y_0 - m_1x_0$. Substituting this into the formula for $f_1(x)$, we have $f_1(x) = m_1x + y_0 - m_1x_0$, which we can rewrite as $f_1(x) = m_1(x - x_0) + y_0$. Similarly, we have $f_2(x) = m_2(x - x_0) + y_0$. From Fig. 2-8, we guess that it is possible to take $I = (-\infty, \infty)$. To finish the proof, we must show that $x < x_0$ implies $f_2(x) < f_1(x)$ and that $x > x_0$ implies $f_2(x) > f_1(x)$.

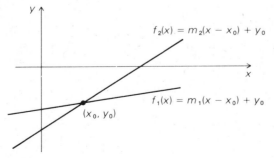

Fig. 2-8 f_1 and f_2 both pass through (x_0, y_0) and f_2 has larger slope.

Assume that $x < x_0$. Then $x - x_0 < 0$. Since $m_2 - m_1 > 0, x - x_0$ and $m_2 - m_1$ have opposite signs so that

$$(m_2 - m_1)(x - x_0) < 0$$

so

$$m_2(x - x_0) - m_1(x - x_0) < 0$$

so

$$m_2(x - x_0) < m_1(x - x_0)$$

and

$$m_2(x - x_0) + y_0 < m_1(x - x_0) + y_0$$

which says exactly that $f_2(x) < f_1(x)$.

If $x > x_0$, we have $x - x_0 > 0$, so $(m_2 - m_1)(x - x_0) > 0$. A chain of manipulations like the one above (which you should write out yourself) leads to the conclusion $f_2(x) > f_1(x)$. That finishes the proof.

Terminology   It is useful to be able to speak of graphs overtaking one another. If G_1 and G_2 are curves in the plane which are the graphs of functions $f_1(x)$ and $f_2(x)$, we will sometimes say that G_2 overtakes G_1 at a point when what is really meant is that f_2 overtakes f_1. Thus, Theorem 3 may be rephrased as follows (see Fig. 2-9):

If the line l_1 with slope m_1 meets the line l_2 with slope m_2 at (x_0, y_0), and if $m_2 > m_1$, then l_2 overtakes l_1 at x_0.

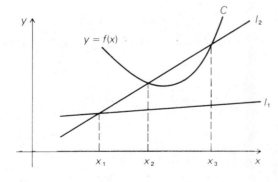

Fig. 2-9 l_2 overtakes l_1 at x_1; l_2 overtakes C at x_2; C overtakes l_2 at x_3.

(This statement only applies to those lines which are graphs of functions—we may not speak of a vertical line overtaking or being overtaken by anything.)

Solved Exercises

4. Where does $1/x$ change sign?

5. Let $f(x) = 3x + 2$, $g(x) = x + 2$. Show in two ways that f overtakes g at 0: (a) by the definition of overtaking; (b) by Theorem 3.

6. Let $f(x) = -3$ and $g(x) = -x^2$. At what point or points does f overtake g? Construct an interval which works. Sketch.

7. Let $f(x) = 2x^2$ and $g(x) = 5x - 3$. Show that g overtakes f at $x = 1$. Is $(-\infty, \frac{5}{4})$ an interval which works? Is it the largest one?

Exercises

5. Describe the change of sign at $x = 0$ of the function $f(x) = mx$ for various values of m. Can you find a transition point on the "m axis" where a certain change takes place?

6. Let $f(x) = x^2 - 2x - 3$ and $g(x) = 2x - 2$. Where does g overtake f? Find an interval which works. Sketch.

7. Let $f(x) = x^3 - x$ and $g(x) = 2x$. At what point or points does f overtake g? Does $(-3, 0)$ work for any of these points? If not, why not? Sketch.

8. Let $f(x) = -x^2 + 4$ and $g(x) = 3x - 2$. At what point or points does f overtake g? Find the largest interval which works. Sketch.

9. Let $f(x) = 1/(1 - x)$ and $g(x) = -x + 1$.

   (a) Show that f overtakes g at $x = 0$. Find an interval which works. Sketch.

   (b) Show that f overtakes g at $x = 2$. Is $(1, 3)$ an interval which works? What is the largest interval which works?

The Derivative

The derivative was defined in Chapter 1. To rephrase that definition using the language of transitions, we shall use the following terminology. Let f be a function whose domain contains an open interval about x_0. Let A be the set of numbers m such that the linear function $f(x_0) + m(x - x_0)$ (whose graph is the line through $(x_0, f(x_0))$ with slope m) is overtaken by f at x_0. Let B be the set of numbers m such that the linear function $f(x_0) + m(x - x_0)$ overtakes f at x_0.

Theorem 4 *A number m_0 is a transition point from A to B if and only if m_0 is the derivative of f at x_0.*

Proof First, suppose that m_0 is a transition point from A to B. Thus there is an open interval I about m_0 such that (i) if $m_1 \in I$ and $m_1 < m_0$, then m_1 is in A but not in B, and (ii) if $m_1 \in I$ and $m_1 > m_0$, then m_1 is in B but not in A. Let $m < m_0$. Choose $m_1 \in I$ such that $m \leqslant m_1 < m_0$. (Why can we do this?) By (i), the function $f(x_0) + m_1(x - x_0)$ is overtaken by f at x_0. By Theorems 2 and 3, $f(x_0) + m(x - x_0)$ is also overtaken by f at x_0; i.e. $f(x) - [f(x_0) + m(x - x_0)]$ changes sign from negative to positive at x_0. This gives condition 1 of the definition of the derivative, and condition 2 is proved in the same way.

Conversely, if m_0 is the derivative of f at x_0, then the definition of the derivative shows that $A = (-\infty, m_0)$ and $B = (m_0, \infty)$, and so m_0 is the transition point from A to B.

In Chapter 1 we proved that the derivative is unique if it exists. In the present terminology, this means that there is at most one transition point from A to B. (In Problems 18-20, the reader is invited to prove this directly from the definition of A and B).

Solved Exercise

8. Let $f(x) = x^2 + \frac{3}{2}x + 2$ and $x_0 = 2$. Construct the sets A and B and use them to calculate $f'(0)$.

Exercises

10. Calculate the derivative of $f(x) = x^2 - x$ at $x = 1$ using Theorem 4.

11. For each of the following functions, find the sets A and B involved in Theorem 4 and show that the derivative does not exist at the specific point:

(a) $f(x) = \begin{cases} x & \text{if } x \leqslant 1 \\ 2x - 1 & \text{if } x > 1 \end{cases}$, $x_0 = 1$

(b) $f(x) = \begin{cases} 2x + 4 & \text{if } x \leqslant -1 \\ 2x + 3 & \text{if } x > -1 \end{cases}$, $x_0 = -1$

(c) $f(x) = \begin{cases} x^2 & \text{if } x \leqslant 0 \\ x & \text{if } x > 0 \end{cases}$, $x_0 = 0$

12. Is the following an acceptable definition of the tangent line? "The tangent line through a point on a graph is the one line which neither overtakes nor is overtaken by the graph." If so, discuss. If not, give an example.

Problems for Chapter 2

1. Describe the following as transition points.
   (a) Turning on a light.   (b) $100°$ Centigrade.
   (c) An aircraft landing.   (d) Signing a contract.

2. What transition points can you identify in the following phenomena?
   (a) Breathing.   (b) A heart beating.
   (c) Blinking.   (d) Walking.
   (e) A formal debate.   (f) Firefighting.
   (g) Marriage.   (h) Receiving exam results.
   (i) Solving homework problems.

3. Let A be the set of areas of *triangles* inscribed in a circle of radius 1, B the set of areas of circumscribed triangles. Is there a transition point from A to B? (You may assume that the largest inscribed and smallest circumscribed triangles are equilateral.) What happens if you use quadrilaterals instead of triangles? Octagons?

4. For each of the following pairs of sets, find the transition points from A to B and from B to A.
   (a) $A = (0, 1)$; $B = (-\infty, \frac{1}{2})$ and $(1, 3)$.
   (b) $A = $ those x for which $x^2 < 3$; $B = $ those x for which $x^2 > 3$.
   (c) $A = $ those x for which $x^3 \leqslant 4$; $B = $ those x for which $x^3 \geqslant 4$.
   (d) $A = $ those a for which the equation $x^2 + a = 0$ has two real roots;
   $B = $ those a for which the equation $x^2 + a = 0$ has less than two real roots.
   (e) $A = $ those x for which $\sqrt{x^2 - 4}$ exists (as a real number):
   $B = $ those x for which $\sqrt{x^2 - 4}$ does not exist (as a real number).

5. For which values of n (positive and negative) does x^n change sign at 0?

6. For each of the following pairs of functions, find:
   1. The set A where $f(x) < g(x)$.
   2. The set B where $f(x) > g(x)$.
   3. The transition points from A to B and from B to A.
   4. Intervals which work for f overtaking g and g overtaking f.

   Make a sketch in each case.
   (a) $f(x) = 7x + 2$; $g(x) = 2x - 4$
   (b) $f(x) = x^2 + 2x + 2$; $g(x) = -x + 3$
   (c) $f(x) = -2x^2 + 4x - 3$; $g(x) = 2x - 5$
   (d) $f(x) = -x^2 + 2x + 1$; $g(x) = x^2 - 1$
   (e) $f(x) = -x^3 + 4x$; $g(x) = x^2 - 2x$
   (f) $f(x) = 1/x^2$ $(x \neq 0)$; $g(x) = -x^2 + 1$
   (g) $f(x) = -1/x^2$ $(x \neq 0)$; $g(x) = [1/(x+2)] - 2$ $(x \neq -2)$

7. For which values of m does $f(x) = m(x - 1) + 1$ overtake $g(x) = x^2$ at 1?

8. Show that f changes sign at x_0 if and only if f overtakes or is overtaken by the zero function $(g(x) = 0$ for all $x)$ at x_0. (This problem and the next one show that the concepts of overtaking and sign change can be defined in terms of one another.)

9. Let f and g be functions, and define h by $h(x) = f(x) - g(x)$, for all x such that $f(x)$ and $g(x)$ are both defined. Prove that f overtakes g at x_0 if and only if h changes sign from negative to positive at x_0.

In Problems 10 to 17, let

$$f(x) = 2x^2 - 5x + 2, \qquad g(x) = \tfrac{3}{4}x^2 + 2x,$$
$$h(x) = -3x^2 + x + 3, \qquad k(x) = 3x - 4, \text{ and}$$
$$l(x) = -2x + 3$$

10. Find the derivative of $f(x) + g(x)$ at $x = 1$.

11. Find the derivative of $3f(x) - 2h(x)$ at $x = 0$.

12. Find the equation of the tangent line to
    (a) $f(x)$ at $x = 1$ (b) $g(x)$ at $x = -2$
    (c) $h(x)$ at $x = 100$ (d) $k(x)$ at $x = -10^8$ (sketch)

13. Where does $l(x)$ overtake $k(x)$?

14. Where does $l(x)$ overtake the tangent line to $h(x)$ at $x = -1$?

15. For what real number c is the line $y = ck(x)$ parallel to the tangent line to $f(x)$ at $x = 2$?

16. Where does $g(x) + l(x)$ overtake $f(x) + k(x)$? What are the derivatives of $g + l$ and $f + k$ at these points?

17. Let $f(t) = 2t^2 - 5t + 2$ be the position of object A, and let $h(t) = -3t^2 + t + 3$ be the position of object B.

    (a) When is A moving faster than B?

    (b) How fast is B going when A stops?

    (c) When does B turn around?

18. A set of numbers S is called *convex* if, whenever x_1 and x_2 lie in S and $x_1 < y < x_2$, then y lies in S too. Prove that the sets A and B defined in the preamble to Theorem 4 are convex.

19. Let A and B be sets of real numbers such that A is convex (see Problem 18). Prove that there is *at most* one point of transition from A to B.

20. Use Problems 18 and 19 to prove the uniqueness of derivatives.

21. (a) Find sets A, B, and C such that there are transition points from A to B, from B to C, and from C to A.

    (b) Prove that this cannot happen if A, B, and C are convex (see Problem 18).

22. Let $f(x)$ and $g(x)$ be functions which are differentiable at x_0.

    (a) Prove that, if $f'(x_0) > g'(x_0)$, then $f(x)$ overtakes $g(x)$ at x_0. [*Hint:* Use a line with slope in the interval $(g'(x_0), f'(x_0))$].

    (b) Prove that, if $f'(x_0) < g'(x_0)$, then $f(x)$ is overtaken by $g(x)$ at x_0.

    (c) Give examples to show that, if $f'(x_0) = g'(x_0)$, then it is possible that $f(x)$ overtakes $g(x)$ at x_0, or $f(x)$ is overtaken by $g(x)$ at x_0, or neither.

    (d) Solve Problem 6 by using (a) and (b).

23. Explain how Fig. 2-10 illustrates the definition of the derivative of $y = x^2$.

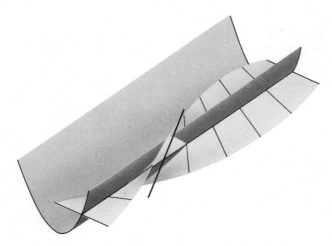

Fig. 2-10 This illustrates the derivative?

3 Algebraic Rules of Differentiation

We shall now prove the sum, constant multiple, product, and quotient rules of differential calculus. (Practice with these rules must be obtained from a standard calculus text.) Our proofs use the concept of "rapidly vanishing functions" which we will develop first. (The reader will be assumed to be familiar with absolute values.)

Rapidly Vanishing Functions

We say that a function f *vanishes at* x_0, or that x_0 is a *root of* f, if $f(x_0) = 0$. If we compare various functions which vanish near a point, we find that some vanish "more rapidly" than others. For example, let $f(x) = x - 2$, and let $g(x) = 20(x - 2)^2$. They both vanish at 2, but we notice that the derivative of g vanishes at 2 as well, while that of f does not. Computing numerically, we find

$f(3)$	$= 1$	$g(3)$	$= 20$
$f(1)$	$= -1$	$g(1)$	$= 20$
$f(2.1)$	$= 0.1$	$g(2.1)$	$= 0.2$
$f(1.9)$	$= -0.1$	$g(1.9)$	$= 0.2$
$f(2.01)$	$= 0.01$	$g(2.01)$	$= 0.002$
$f(1.99)$	$= -0.01$	$g(1.99)$	$= 0.002$
$f(2.001)$	$= 0.001$	$g(2.001)$	$= 0.00002$
$f(1.999)$	$= -0.001$	$g(1.999)$	$= 0.00002$
$f(2.0001)$	$= 0.0001$	$g(2.0001)$	$= 0.0000002$
$f(1.9999)$	$= -0.0001$	$g(1.9999)$	$= 0.0000002$

As x approaches 2, $g(x)$ appears to be dwindling away more rapidly than $f(x)$. Guided by this example, we make the following definition.

> **Definition** We say that a function $r(x)$ *vanishes rapidly* at x_0 if $r(x_0) = 0$ and $r'(x_0) = 0$.

The following theorem shows that rapidly vanishing functions, as we have defined them, really do vanish quickly. This theorem will be useful in proving properties of rapidly vanishing functions, as well as in establishing the connection between transitions and limits (see Chapter 13).

Theorem 1 *Let r be a function such that $r(x_0) = 0$. Then $r(x)$ vanishes rapidly at $x = x_0$ if and only if, for every positive number ϵ, there is an open interval I about x_0 such that, for all $x \neq x_0$ in I, $|r(x)| < \epsilon|x - x_0|$.*

Proof The theorem has two parts. First of all, suppose that $r(x)$ vanishes rapidly at x_0, and that ϵ is a positive number. We will find an interval I with the required properties. Since $-\epsilon < r'(x_0) = 0 < \epsilon$, the line through $(x_0, r(x_0))$ with slope $-\epsilon$ [slope ϵ] is overtaken by [overtakes] the graph of f at x_0. Since $r(x_0) = 0$, the equations of these two lines are $y = -\epsilon(x - x_0)$ and $y = \epsilon(x - x_0)$. (Refer to Fig. 3-1.)

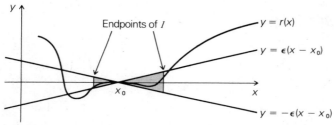

Fig. 3-1 The graph $y = r(x)$ is squeezed between the lines $y = -\epsilon(x - x_0)$ and $y = \epsilon(x - x_0)$.

Let I_1 be an interval which works for $r(x)$ overtaking $-\epsilon(x - x_0)$ at x_0, and I_2 an interval which works for $\epsilon(x - x_0)$ overtaking $r(x)$ at x_0. Choose an open interval I containing x_0 and contained in both I_1 and I_2.

For $x < x_0$ and x in I, we have $r(x) < -\epsilon(x - x_0)$, since $r(x)$ overtakes $-\epsilon(x - x_0)$, and $r(x) > \epsilon(x - x_0)$, since $\epsilon(x - x_0)$ overtakes $r(x)$. We may rewrite these two inequalities as $-\epsilon(x_0 - x) < r(x) < \epsilon(x_0 - x)$, i.e., $-\epsilon|x - x_0| < r(x) < \epsilon|x - x_0|$, since $|x - x_0| = x_0 - x$ when $x < x_0$.

Now we assume $x > x_0$ in I. Our overtakings imply that $r(x) > -\epsilon(x - x_0)$ and $r(x) < \epsilon(x - x_0)$, so $-\epsilon(x - x_0) < r(x) < \epsilon(x - x_0)$, and, once again, $-\epsilon|x - x_0| < r(x) < \epsilon|x - x_0|$.

We have shown that, for $x \neq x_0$ and x in I, $-\epsilon|x - x_0| < r(x) < \epsilon|x - x_0|$. But this is the same as $|r(x)| < \epsilon|x - x_0|$. We have finished half of our "if and only if" proof. Geometrically speaking, we have shown that the graph of $r(x)$ for $x \in I$ lies inside the shaded "bow-tie" region in Fig. 3-1.

For the second half of the proof, we assume that, for any $\epsilon > 0$, there exists an interval I such that, for all x in I, $x \neq x_0$, $|r(x)| < \epsilon|x - x_0|$.

Reversing the steps in the preceding argument shows that, for all $\epsilon > 0$, the line $y = -\epsilon(x - x_0)$ is overtaken by the graph of $r(x)$ at x_0, while the line $y = \epsilon(x - x_0)$ overtakes the graph. Thus the set A in the definition of the derivative contains all negative numbers, and B contains all positive numbers; therefore, 0 is a point of transition from A to B. Thus $r'(x_0) = 0$, so r vanishes rapidly at x_0.

The next theorem shows the importance of rapidly vanishing functions in the study of differentiation.

Theorem 2 *A function f is differentiable at x_0 and $m_0 = f'(x_0)$ if and only if the function r(x), defined by*

$$r(x) = f(x) - [f(x_0) + m_0(x - x_0)]$$

vanishes rapidly at x_0.

The function $r(x)$ represents the *error,* or the *remainder* involved in approximating f by its tangent line at x_0. Another way of stating Theorem 2 is that f is differentiable at x_0, with derivative $f'(x_0) = m_0$, if and only if f can be written as a sum $f(x) = f(x_0) + m_0(x - x_0) + r(x)$, where $r(x)$ vanishes rapidly at x_0. Once we know how to recognize rapidly vanishing functions, Theorem 2 will provide a useful test for differentiability and a tool for computing derivatives.

Notice that Theorem 2, like Theorem 1, is of the "if and only if" type. Thus, it has two independent parts. We must prove that if $f(x)$ is differentiable at x_0 and $m_0 = f'(x_0)$, then $r(x)$ vanishes rapidly at x_0, and we must prove as well that if $r(x)$ vanishes rapidly at x_0, then $f(x)$ is differentiable at x_0 and $m_0 = f'(x_0)$. (Each of the two parts of the theorem is called the *converse* of the other.) In the proof which follows, we prove the second part first, since it is convenient to do so.

Proof of Theorem 2 Suppose that $r(x) = f(x) - f(x_0) - m_0(x - x_0)$ vanishes rapidly at x_0. We will show that $f'(x_0) = m_0$. We first show that, for $m > m_0$, the line through $(x_0, f(x_0))$ overtakes the graph of f at x_0. (Refer to Fig. 3-2 as you read the following argument.) To begin, note that if $m > m_0$, then $m - m_0 > 0$. Since $r'(x_0) = 0$, the line through $(x_0, r(x_0))$ with slope $m - m_0$ overtakes the graph of r at x_0. Since $r(x_0) = 0$, the equation of this line is $y = (m - m_0)(x - x_0)$. By the definition of overtaking, there is an open interval I around x_0 such that (i) $x < x_0$ in I implies $(m - m_0)(x - x_0) < r(x)$, i.e.,

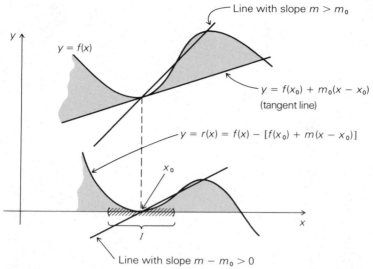

Fig. 3-2 Illustrating the proof of Theorem 2.

$$(m - m_0)(x - x_0) < f(x) - f(x_0) - m_0(x - x_0) \qquad (A_1)$$

and (ii) $x > x_0$ in I implies

$$(m - m_0)(x - x_0) > f(x) - f(x_0) - m_0(x - x_0) \qquad (A_2)$$

Adding $f(x_0) + m_0(x - x_0)$ to both sides of (A_1) and (A_2) gives

$$f(x_0) + m(x - x_0) < f(x), \quad \text{for } x < x_0 \text{ in } I$$

and

$$f(x_0) + m(x - x_0) > f(x), \quad \text{for } x > x_0 \text{ in } I$$

In other words, the line $y = f(x_0) + m(x - x_0)$ overtakes the graph of f at x_0. Recall that m was any number greater than m_0.

Similarly, one proves that, for $m < m_0$, the line $y = f(x_0) + m(x - x_0)$ is overtaken by the graph of f at x_0, so f is differentiable at x_0, and m_0 must be the derivative of f at x_0 by the definition on p. 6 and Theorem 4, p. 27. This completes the proof of half of Theorem 2.

Next, we assume that f is differentiable at x_0 and that $m_0 = f'(x_0)$. Reversing the steps above shows that the line $y = c(x - x_0)$ overtakes the graph of r for $c > 0$ and is overtaken by it for $c < 0$. Clearly, $r(x_0) = 0$, so r vanishes rapidly at x_0.

Solved Exercises*

1. Let $g_1(x) = x$, $g_2(x) = x^2$, $g_3(x) = x^3$. Compute their values at $x = 0.1$, $0.01, 0.002$, and 0.0004. Discuss.

2. Prove:

   (a) If f and g vanish at x_0, so does $f + g$.

   (b) If f vanishes at x_0, and g is any function which is defined at x_0, then fg vanishes at x_0.

3. Prove that $(x - x_0)^2$ vanishes rapidly at x_0.

Exercises

1. Fill in the details of the last two paragraphs of the proof of Theorem 2.

2. Do Solved Exercise 3 by using Theorem 1.

3. Prove that $f(x) = x^3$ is rapidly vanishing at $x = 0$.

4. Let $g(x)$ be a quadratic polynomial such that $g(0) = 5$.

   (a) Can $g(x)$ vanish rapidly at some integer?

   (b) Can $g(x)$ vanish rapidly at more than one point?

5. The polynomials $x^3 - 5x^2 + 8x - 4$, $x^3 - 4x^2 + 5x - 2$, and $x^3 - 3x^2 + 3x - 1$ all vanish at $x_0 = 1$. By evaluating these polynomials for values of x very near 1, on a calculator, try to guess which of the polynomials vanish rapidly at 1. (Factoring the polynomials may help you to understand what is happening.)

The Sum and Constant Multiple Rules

The sum rule states that $(f + g)'(x) = f'(x) + g'(x)$. To prove this, we must show that the remainder for $f + g$, namely

$$\{f(x) + g(x)\} - [\{f(x_0) + g(x_0)\} - \{f'(x_0) + g'(x_0)\}(x - x_0)]$$

vanishes rapidly at x_0, for then Theorem 2 would imply that $f(x) + g(x)$ is dif-

---

*Solutions appear in the Appendix.

ferentiable at x_0 with derivative $m_0 = f'(x_0) + g'(x_0)$. We may rewrite the remainder as

$$[f(x) - f(x_0) - f'(x_0)(x - x_0)] + [g(x) - g(x_0) - g'(x_0)(x - x_0)].$$

By Theorem 2, each of the expressions in square brackets represents a function which vanishes rapidly at x_0, so we need to show that the sum of two rapidly vanishing functions is rapidly vanishing.

The constant multiple rule states that $(af)'(x) = af'(x)$. The proof of this rests upon the fact that a constant multiple of a rapidly vanishing function is again rapidly vanishing. Theorem 3 proves these two basic properties of rapidly vanishing functions.

Theorem 3

1. *If $r_1(x)$ and $r_2(x)$ vanish rapidly at x_0, then so does $r_1(x) + r_2(x)$.*

2. *If $r_1(x)$ vanishes rapidly at x_0, and a is any real number, then $ar_1(x)$ vanishes rapidly at x_0.*

Proof 1. Let $r(x) = r_1(x) + r_2(x)$, where $r_1(x)$ and $r_2(x)$ vanish rapidly at x_0. By Solved Exercise 2, $r(x_0) = 0$; we will use Theorem 1 to show that $r(x)$ vanishes rapidly at x_0. Given $\epsilon > 0$, we must find an interval I such that $|r(x)| < \epsilon |x - x_0|$ for all $x \neq x_0$ in I. Since $r_1(x)$ and $r_2(x)$ vanish rapidly at x_0, there are intervals I_1 and I_2 about x_0 such that $x \neq x_0$ in I_1 implies that $|r_1(x)| < (\epsilon/2) |x - x_0|$, while $x \neq x_0$ in I_2 implies $|r_2(x)| < (\epsilon/2) |x - x_0|$. (We can apply Theorem 1 with any positive number, including $\epsilon/2$, in place of ϵ.) Let I be an interval containing x_0 and contained in both I_1 and I_2. For $x \neq x_0$ in I, we have both inequalities: $|r_1(x)| < (\epsilon/2) |x - x_0|$ and $|r_2(x)| < (\epsilon/2) |x - x_0|$. Adding these inequalities gives

$$|r_1(x)| + |r_2(x)| < \frac{\epsilon}{2} |x - x_0| + \frac{\epsilon}{2} |x - x_0| = \epsilon |x - x_0|$$

(You should now be able to see why we used $\epsilon/2$.) The triangle inequality for absolute values states that $|r_1(x) + r_2(x)| \leqslant |r_1(x)| + |r_2(x)|$, so we have $|r(x)| < \epsilon |x - x_0|$ for $x \neq x_0$ in I.

2. Let $r(x) = ar_1(x)$. By Solved Exercise 2, $r(x_0) = 0$. Given $\epsilon > 0$, we apply Theorem 1 to r_1 vanishing rapidly at x_0 to obtain an interval I about x_0 such that $|r_1(x)| < (\epsilon/|a|) |x - x_0|$ for $x \neq x_0$ in I. (If $a = 0$, $ar(x) = 0$ is obviously rapidly vanishing, so we need only deal with the case $a \neq 0$.) Now we have, for $x \neq x_0$ in I,

$$|r(x)| = |ar_1(x)| = |a| \, |r_1(x)| < |a| \, \frac{\epsilon}{|a|} \, |x - x_0| = \epsilon|x - x_0|$$

or $|r(x)| < \epsilon|x - x_0|$, and we are done.

Theorem 4

1. *(Sum Rule)*. *If the functions $f(x)$ and $g(x)$ are differentiable at x_0, then so is the function $f(x) + g(x)$, and its derivative at x_0 is $f'(x_0) + g'(x_0)$.*

2. *(Constant Multiple Rule)*. *If $f(x)$ is differentiable at x_0, and a is any real number, then the function $af(x)$ is differentiable at x_0, and its derivative there is $af'(x_0)$.*

Proof   1. Since $f(x)$ and $g(x)$ are differentiable at x_0, Theorem 2 tells us that

$$r_1(x) = f(x) - f(x_0) - f'(x_0)(x - x_0)$$

and

$$r_2(x) = g(x) - g(x_0) - g'(x_0)(x - x_0)$$

vanish rapidly at x_0. Adding these two equations, we conclude by Theorem 3 that

$$r_1(x) + r_2(x) = [f(x) + g(x)] - [f(x_0) + g(x_0)]$$
$$- [f'(x_0) + g'(x_0)] (x - x_0)$$

is also rapidly vanishing at x_0. Hence, by Theorem 2, $f(x) + g(x)$ is differentiable at x_0 with derivative equal to $f'(x_0) + g'(x_0)$.

2. See Solved Exercise 5.

Solved Exercises

4. Prove that, if $r_1(x)$ and $r_2(x)$ vanish rapidly at x_0, then so does $r_1(x) - r_2(x)$.

5. Prove part (2) of Theorem 4 (the constant multiple rule).

Exercises

6. Let a be a nonzero constant, and assume that af overtakes g at x_0. Prove that f overtakes $(1/a) g$ at x_0 if $a > 0$, while f is overtaken by $(1/a) g$ at x_0 if $a < 0$.

7. Show that, if you *assume* the sum and constant multiple rules, then Theorem 3 is an easy consequence. (This means that Theorem 3 is a *special case* of the sum and constant multiple rules. Our proof of Theorem 4 proceeded from this special case to the general case.)

8. Prove that, if $f_1(x)$ overtakes $g_1(x)$ at x_0, and $f_2(x)$ overtakes $g_2(x)$ at x_0, then $f_1(x) + f_2(x)$ overtakes $g_1(x) + g_2(x)$ at x_0.

9. If $f_1(x) + f_2(x)$ is differentiable at x_0, are $f_1(x)$ and $f_2(x)$ necessarily differentiable there? Can just one of them be nondifferentiable at x_0?

10. Show, by calculating the derivative, that the quadratic polynomial $ax^2 + bx + c$ vanishes rapidly at x_0 if and only if x_0 is a *double root* of the equation $ax^2 + bx + c = 0$. What does the graph look like in this case?

The Product Rule

The sum rule depended on the fact that the sum of two rapidly vanishing functions is again rapidly vanishing. For the product rule, we need a similar result for products, where only one factor is known to be rapidly vanishing.

Theorem 5 *If $r(x)$ vanishes rapidly at x_0 and $f(x)$ is differentiable at x_0, then $f(x)r(x)$ vanishes rapidly at x_0.*

Proof Note that part 2 of Theorem 3 is a special case of this theorem, where $f(x)$ is constant. We prove Theorem 5 in two steps, the first of which shows that $f(x)$ can be "sandwiched" between two constant values.

  Step 1 We will prove that there is a constant $B > 0$ and an interval I_1 about x_0 such that, for x in I_1, we have

$$-B < f(x) < B, \quad \text{i.e.,} \quad |f(x)| < B.$$

(Refer to Fig. 3-3.) The number B is called a *bound* for $|f(x)|$ near x_0. Through the point $(x_0, f(x_0))$, we draw the two lines with slope $f'(x_0) + 1$ and $f'(x_0) - 1$. The first of them overtakes the graph of f at x_0; the second

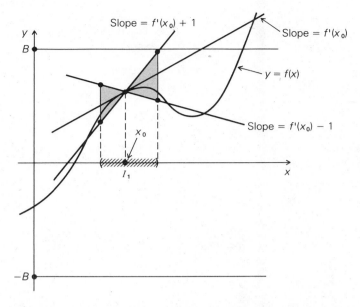

Slope = $f'(x_0) + 1$

Slope = $f'(x_0)$

$y = f(x)$

Slope = $f'(x_0) - 1$

Fig. 3-3 The graph of f on the interval I_1 lies between the lines $y = -B$ and $y = B$.

is overtaken. (We could just as well have used slopes $f'(x_0) + \frac{1}{2}$ and $f'(x_0) - \frac{1}{2}$, or any other numbers bracketing $f'(x_0)$.) On a sufficiently small interval I_1, the graph of f lies between these two lines. (I_1 is any interval contained in intervals which work for both overtakings.) Now choose B large enough so that the bow-tie region between the lines and above I_1 lies between the lines $y = -B$ and $y = B$. The reader may fill in the algebra required to determine a possible choice for B. (See Solved Exercise 6.)

   Step 2.   Clearly $f(x)r(x)$ vanishes at x_0. We now apply Theorem 1 just as we did in proving part (ii) of Theorem 3.

   Given $\epsilon > 0$, since $r(x)$ vanishes rapidly at x_0, we can find an interval I about x_0 such that, for $x \neq x_0$ in I, $|r(x)| < (\epsilon/B)\,|x - x_0|$, where B is the bound from step 1. Now we have, for $x \neq x_0$ in I, $|f(x)r(x)| = |f(x)|\,|r(x)| < B\,(\epsilon/B)\,|x - x_0| = \epsilon\,|x - x_0|$. Thus, by Theorem 1, $f(x)r(x)$ vanishes rapidly at x_0.

We can now deduce the product rule from Theorem 5 by a computation.

Theorem 6   Product Rule. *If the functions $f(x)$ and $g(x)$ are differentiable at x_0, then so is the function $f(x)g(x)$, and its derivative at x_0 is*

$$f'(x_0)g(x_0) + f(x_0)g'(x_0)$$

Proof By Theorem 2, $f(x) = f(x_0) + f'(x_0)(x - x_0) + r_1(x)$ and $g(x) = g(x_0) + g'(x_0)(x - x_0) + r_2(x)$ where $r_1(x)$ and $r_2(x)$ vanish rapidly at x_0. Multiplying the two expressions gives

$$
\begin{aligned}
f(x)g(x) &= f(x_0)[g(x_0) + g'(x_0)(x - x_0) + r_2(x)] \\
&\quad + f'(x_0)(x - x_0)[g(x_0) + g'(x_0)(x - x_0) + r_2(x)] \\
&\quad + r_1(x)[g(x_0) + g'(x_0)(x - x_0) + r_2(x)] \\
&= f(x_0)g(x_0) + [f'(x_0)g(x_0) + f(x_0)g'(x_0)](x - x_0) \\
&\quad + f(x_0)r_2(x) + r_1(x)g(x_0) + f'(x_0)g'(x_0)(x - x_0)^2 \\
&\quad + f'(x_0)(x - x_0)r_2(x) + r_1(x)g'(x_0)(x - x_0) + r_1(x)r_2(x)
\end{aligned}
$$

Now each of the last two lines in the preceding sum vanishes rapidly at x_0: $f(x_0)r_2(x) + r_1(x)g(x_0)$ is a sum of constant multiples of rapidly vanishing functions (apply Theorem 3); $f'(x_0)g'(x_0)(x - x_0)^2$ is a constant multiple of $(x - x_0)^2$, which is rapidly vanishing by Solved Exercise 3; each of the next two terms is the product of a linear function and a rapidly vanishing function (apply Theorem 4), and the last term is the product of two rapidly vanishing functions (apply Theorem 4 again). Applying Theorem 3 to the sum of the last two lines, we conclude that

$$f(x)g(x) = f(x_0)g(x_0) + [f'(x_0)g(x_0) + f(x_0)g'(x_0)](x - x_0) + r(x),$$

where $r(x)$ vanishes rapidly at x_0. Theorem 2 therefore shows that $f(x)g(x)$ is differentiable at x_0 with derivative $f'(x_0)g(x_0) + f(x_0)g'(x_0)$.

Note that the correct formula for the derivative of a product appeared as the coefficient of $(x - x_0)$ in our computation; there was no need to know it in advance.

Now that we have proven the product rule, we may use one of its important consequences: the derivative of x^n is nx^{n-1} (see Solved Exercise 7).

Solved Exercises

6. In step 1 of the proof above, estimate how large B must be so that $-B < f(x) < B$ for all x in I.

7. Prove that $f(x) = x^4$ vanishes rapidly at 0.

8. Find a function which vanishes rapidly at both 3 and 7.

Exercises

11. Find a function which vanishes rapidly at 1, 2, and 3. Sketch a graph of this function.

12. Show that $(x - a)^n$ vanishes rapidly at a if n is any positive integer greater than or equal to 2.

13. Recall that, if $g(x)$ is a polynomial and $g(a) = 0$, then $g(x) = (x - a)h(x)$, where $h(x)$ is another polynomial. We call a a *multiple root* of $g(x)$ if $h(a) = 0$. Prove that a is a multiple root of $g(x)$ if and only if $g(x)$ vanishes rapidly at a.

14. If $f(x)$ and $g(x)$ are defined in an interval about x_0, and $f'(x_0)$ and $(fg)'(x_0)$ both exist, does $g'(x_0)$ necessarily exist? (Compare Exercise 9.)

The Quotient Rule

Theorem 7 Suppose that g is differentiable at x_0 and that $g(x_0) \neq 0$. Then there is an interval I about x_0 on which $g(x)$ is never zero, and $1/g$ is differentiable at x_0 with derivative $-g'(x_0)/g(x_0)^2$.

Proof Suppose $g(x_0) > 0$. (The case $g(x_0) < 0$ is discussed at the end of this proof.) Then a bow-tie argument like that on p. 39 shows that $g(x) > \frac{1}{2} g(x_0)$ for x in some interval I about x_0. (See Fig. 3-4 and Solved Exercise 9.) Since $0 < \frac{1}{2} g(x_0) < g(x)$ for x in I, we have $0 < 1/g(x) < 2/g(x_0)$ for x in I. In Theorem 5, we actually proved that the product of a rapidly vanishing function and a function bounded between two values is again rapidly vanishing. By Theorem 2, we must show that

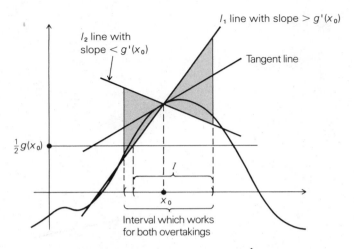

Fig. 3-4 $g(x)$ is bounded below near x_0 by $\frac{1}{2}g(x_0)$.

$$r(x) = \frac{1}{g(x)} - \frac{1}{g(x_0)} + \frac{g'(x_0)}{g(x_0)^2}(x - x_0)$$

vanishes rapidly at x_0. Collecting terms over a common denominator, we get

$$r(x) = \frac{1}{g(x)g(x_0)^2}\{g(x_0)^2 - g(x)g(x_0) + g'(x_0)g(x)(x - x_0)\}$$

Since $1/[g(x)g(x_0)^2]$ lies between the values 0 and $[2/g(x_0)]\,[1/g(x_0)^2]$, it suffices to prove that the expression in braces,

$$r_1(x) = g(x_0)^2 - g(x)g(x_0) + g'(x_0)g(x)(x - x_0)$$

vanishes rapidly at zero. Setting $x = x_0$ in $r_1(x)$, we get $r_1(x_0) = 0$. We must show that $r_1'(x_0) = 0$. However, now that we have proven the sum, constant multiple, and product rules, we may use them:

$$r_1'(x) = -g'(x)g(x_0) + g'(x_0)g'(x)(x - x_0) + g'(x_0)g(x)$$

wherever g is differentiable. Setting $x = x_0$, we get $r_1'(x_0) = 0$, so $r_1(x)$ vanishes rapidly at x_0, and so does $r(x)$. That finishes the proof for the case $g(x_0) > 0$.

If $g(x_0) < 0$, we can apply the previous argument to $-g$. Namely, write

$$\frac{1}{g(x)} = -\frac{1}{-g(x)}$$

Since $g(x)$ is differentiable at x_0, so is $-g(x)$. But $-g(x_0)$ is positive, so the argument above implies that $1/[-g(x)]$ is differentiable at x_0. Now the constant multiple rule gives the differentiability of $-1/[-g(x)]$.

Corollary   Quotient Rule. *Suppose that f and g are differentiable at x_0 and that $g(x_0) \neq 0$. Then f/g is differentiable at x_0 with derivative*

$$\frac{g(x_0)f'(x_0) - f(x_0)g'(x_0)}{[g(x_0)]^2}$$

Proof   Write $f/g = f \cdot (1/g)$ and apply the product rule and Theorem 7.

With these basic results in hand we can now readily differentiate any rational function. (See your regular calculus text.)

Solved Exercises

9. Using Fig. 3-4, locate the left-hand endpoint of the interval I.

10. Find a function g such that $g(0) \neq 0$, but $g(x) = 0$ for some point x in every interval about 0. Could such a g be differentiable at 0?

Exercises

15. Let a and b be real numbers. Suppose that $f(x)$ vanishes rapidly at a and that $g(x)$ is differentiable at a. Find a necessary and sufficient condition for $f(x)/(g(x) - b)$ to vanish rapidly at a.

16. Could f/g be differentiable at x_0 without f and g themselves being differentiable there?

Problems for Chapter 3 ▬▬▬▬▬▬▬▬▬▬▬▬▬▬

1. Let

$$f(x) = \begin{cases} x^2 & \text{if } x \geqslant 0 \\ -x^2 & \text{if } x \leqslant 0 \end{cases}$$

Does $f(x)$ vanish rapidly at $x = 0$?

2. Find a function $g(x)$ defined for all x such that $x^2 g(x)$ does not vanish rapidly at $x = 0$. Sketch the graphs of $g(x)$ and $x^2 g(x)$.

3. A function f is called *locally bounded* at x_0 if there is an open interval I about x_0 and a constant B such that $|f(x)| \leqslant B$ for x in I. Prove that if r is rapidly vanishing at x_0 and f is locally bounded at x_0, then fr vanishes rapidly at x_0.

4. Let

$$f(x) = \begin{cases} x^2 & \text{if } x \text{ is irrational} \\ -x^2 & \text{if } x \text{ is rational} \end{cases}$$

Show that $f(x)$ is not differentiable at any nonzero value of x_0. Is $f(x)$ differentiable at 0? (Prove your answer.)

5. Suppose that $f(x)$ and $g(x)$ are differentiable at x_0 and that they both vanish there. Prove that their product vanishes rapidly at x_0. [*Hint*: Use the product rule.]

6. Find functions $f(x)$ and $g(x)$, defined for all x, such that $f(0) = g(0) = 0$, but $f(x)g(x)$ does not vanish rapidly at x_0.

7. Suppose that $f(x)$ and $g(x)$ are differentiable at x_0 and that $f(x_0) < g(x_0)$. Prove that there is an interval I about x_0 such that $f(x) < g(x)$ for all x in I. [*Hint*: Consider $g(x) - f(x)$ and look at the proof of Theorem 7.]

8. Prove that, if $r_1(x)$ and $r_1(x) + r_2(x)$ vanish rapidly at x_0, then so does $r_2(x)$.

9. Prove that, if $f(x)g(x)$ vanishes rapidly at x_0, and $g(x)$ is differentiable at x_0 with $g(x_0) \neq 0$, then $f(x)$ vanishes rapidly at x_0. What happens if $g(x_0) = 0$?

- 10. Find a function which vanishes rapidly at every integer.

4 The Real Numbers

An axiomatic treatment of the real numbers provides a firm basis for our reasoning, and it gives us a framework for studying some subtle questions concerning irrational numbers.

To such questions as, "how do we know that there is a number whose square is 2?" and "how is π constructed?" it is tempting to give geometric answers: "$\sqrt{2}$ is the length of the diagonal of square whose side has length 1"; "π is the ratio of the circumference of a circle to its diameter." Answers like this are unsatisfactory, though, since they rely too much on intuition. How does one *define* the circumference of a circle? These questions have been of serious concern to mathematicians for centuries. One way to settle them, without recourse to geometric intuition, is to write down a list of unambiguous rules or *axioms* which enable us to prove all we want.

A set of axioms for the real numbers was developed in the middle part of the 19th Century. These particular axioms have proven their worth without much doubt,* and we will take them for our starting point. In more advanced courses one has to face the question of showing that there exists a system of numbers obeying these axioms, but we shall merely assume this here.

Mathematics that is useful in applications to science is rarely discovered by means of axiom systems. Axiomatics is more frequently the final product of a piece of mathematics created for some need. The axioms for real numbers were agreed on only after centuries of trial and error, and only after the basic theorems were already discovered.

It is *not* our intent to show that *all* the usual manipulative rules follow from the axioms, since that job is too long and is done in algebra courses. Our aim is merely to set out our assumptions in a clear fashion and to give a few illustrations of how to use them.

Addition and Multiplication Axioms

Our first axioms pertain to the operation of addition.

---

*There is still some controversy remaining. Some mathematicians prefer a "constructive" or "intuitionistic" approach; see Heyting, *Intuitionism, An Introduction*, North-Holland (1956), or Bishop, *Foundations of Constructive Analysis*, McGraw-Hill (1967).

I. Addition Axioms   There is an addition operation "+" which associates to every two real numbers x and y a real number $x + y$ called the *sum* of x and y such that:

1. For all x and y, $x + y = y + x$ [commutativity].

2. For all $x, y,$ and z, $x + (y + z) = (x + y) + z$ [associativity].

3. There is a number 0 ("zero") such that, for all x, $x + 0 = x$ [existence of additive identity].

4. For each x, there is a number $-x$ such that $x + (-x) = 0$ [existence of additive inverses].

On the basis of addition axiom (4), we can define the operation of subtraction by $x - y = x + (-y)$.

The next axioms pertain to multiplication and its relation with addition.

II. Multiplication Axioms   There is a multiplication operation "·" which associates to every two real numbers x and y a real number $x \cdot y$, called the *product* of x and y, such that:

1. For all x and y, $x \cdot y = y \cdot x$ [commutativity].

2. For all $x, y,$ and z, $x \cdot (y \cdot z) = (x \cdot y) \cdot z$ [associativity].

3. There is a number 1, which is different from 0, such that, for all x, $x \cdot 1 = x$ [existence of multiplicative identity].

4. For each $x \neq 0$, there is a number $1/x$ such that

$$\left(\frac{1}{x} \cdot x \right) = 1$$

[existence of multiplicative inverses].

5. For all $x, y,$ and z, $x \cdot (y + z) = x \cdot y + x \cdot z$ [distributivity].

On the basis of multiplication axiom 4, we can define the operation of division by

$$\frac{x}{y} = x \cdot \left(\frac{1}{y} \right) \quad \text{(if } y \neq 0)$$

One often writes xy for $x \cdot y$.

Implicit in our rules for addition and multiplication is that the numbers $x + y$ and xy are uniquely specified once x and y are given. Thus we have, for example, the usual rule of algebraic manipulation, "if $x = z$, then $xy = zy$." In fact, $x = z$ means that x and z are the same number, so that multiplying each of them by y must give the same result.

In principle, one could prove all the usual rules of algebraic manipulation from the axioms above, but we will content ourselves with the few samples given in the exercises below.

A final remark: we can define $2 = 1 + 1$, $3 = 2 + 1$, ..., and via division obtain the fractions. As usual, we write x^2 for $x \cdot x$, x^3 for $x^2 \cdot x$, etc.

Solved Exercises*

1. Prove that $x \cdot 0 = 0$ by multiplying the equality $0 + 0 = 0$ by x.

2. Prove that $(x + y)^2 = x^2 + 2xy + y^2$.

3. Prove that $2 \cdot 3 = 6$.

4. Prove that $(-x) \cdot y = -(xy)$.

5. Prove that $\dfrac{a}{b} + \dfrac{c}{d} = \dfrac{ad + bc}{bd}$ if $b \neq 0$ and $d \neq 0$.

Exercises

Prove the following identities.

1. $(-x)(-y) = xy$

2. $(x - y)^2 = x^2 - 2xy + y^2$

3. $(x + y)(x - y) = x^2 - y^2$

4. $\left(\dfrac{a}{b}\right) \cdot \left(\dfrac{c}{d}\right) = \dfrac{ac}{bd}$ $(b \neq 0, d \neq 0)$

5. $\dfrac{1}{(a/b)} = \dfrac{b}{a}$ $(a \neq 0, b \neq 0)$

6. $(a + b)\left(\dfrac{1}{a} + \dfrac{1}{b}\right) = \dfrac{a}{b} + \dfrac{b}{a} + 2$ $(a \neq 0, b \neq 0)$

7. $(-x)\left(\dfrac{1}{x}\right) = -1$ $(x \neq 0)$

---

*Solutions appear in the Appendix.

8. $\frac{1}{2} + \frac{1}{3} = \frac{5}{6}$

9. $a + 1 \neq a$

10. If $a + b = a + c$, then $b = c$.

Order Axioms

From now on we will use, without further justification, the usual rules for algebraic manipulations. The previous exercises were intended to convince the reader that these rules can all be derived from the addition and multiplication axioms. We turn now to the order axioms.

III. Order Axioms There is a relation "$\leqslant$" such that, for certain pairs x and y of real numbers the statement "$x \leqslant y$" (read "x is less than or equal to y") is true. This relation has the following properties:

1. If $x \leqslant y$ and $y \leqslant z$, then $x \leqslant z$ [transitivity].

2. If $x = y$, then $x \leqslant y$ [reflexivity].

3. If $x \leqslant y$ and $y \leqslant x$, then $x = y$ [asymmetry].

4. For any numbers x and y, either $x \leqslant y$ or $y \leqslant x$ is true [comparability].

5. If $x \leqslant y$, and z is any number, then $x + z \leqslant y + z$.

6. If $0 \leqslant x$ and $0 \leqslant y$, then $0 \leqslant xy$.

We write $x < y$ (x is strictly less than y) if $x \leqslant y$ and $x \neq y$. Also, we write $y \geqslant x$ (y is greater than or equal to x) if $x \leqslant y$, and $y > x$ (y is strictly greater than x) if $x < y$. Again, one can prove all the usual properties of the inequality signs from the axioms above. As before, we limit ourselves to a few instances.

Solved Exercises

6. Prove that $0 < 1$.

7. Prove: if $x \leqslant y$ and $c \leqslant 0$, then $cx \geqslant cy$.

8. Let a and b be numbers such that, for any number c with $c < a$, we must have $c < b$. Prove that $a \leqslant b$.

9. Prove: $x^2 \geqslant 0$ for all x.

Exercises

Prove the following statements.

11. $1 < 2$

12. If $a > 1$, then $1/a < 1$.

13. If $a^2 < a$, then $0 < a < 1$.

14. If $a > 0$ and $b < 0$, then $(a + b)^2 < (a - b)^2$.

15. If a and b have the same sign, then $(a + b)^2 > (a - b)^2$.

16. If $x < y$, and z is any number, then $x - z < y - z$.

17. If $a < b$ and $c < d$, then $a + c < b + d$.

18. If a is less than every positive real number, then $a \leqslant 0$.

19. A number a was called *infinitesimal* by the founders of calculus if: (1) $a > 0$; and (2) a is less than every positive real number. Prove that there are no infinitesimal real numbers.*

The Completeness Axiom

We need one more axiom to guarantee that irrational numbers exist. In fact, the rational numbers (i.e., quotients of integers) satisfy all of the axioms in I, II, and III.

To motivate the last axiom, we consider the problem of defining $\sqrt{2}$. Consider the set S consisting of all numbers such that $0 \leqslant x$ and $x^2 \leqslant 2$. If x_1 and x_2 are elements of S and y is a number between them, i.e., $x_1 < y < x_2$, then the order axioms imply that $0 \leqslant y$ (since $0 \leqslant x_1$) and also that $y^2 \leqslant 2$ (since $0 \leqslant y \leqslant x_2$), so that y is again in S. Thus, the set S has "no holes" in the sense that it contains every number between any two of its members. Our intuition tells us that S ought to be an interval. If we could be sure that this were so, we could look at the right-hand endpoint of S (which cannot be ∞, since $1 \in S$ but $2 \notin S$) and establish that $c^2 = 2$. (This is done in Solved Exercise 11.) We would then have found the square root of 2. The completeness axiom makes our intuitive notion into a property of real numbers, taking its place alongside the addition, multiplication, and order axioms. We need one definition before stating the axiom.

---

*There is a modern theory of infinitesimals, but they are not real numbers. The theory is called "nonstandard analysis." (See Keisler, H., *Elementary Calculus:* Prindle, Weber, and Schmidt, Boston (1976).)

Definition A set S of real numbers is *convex* if, whenever x_1 and x_2 belong to S and y is a number such that $x_1 < y < x_2$, then y belongs to S as well.

Any interval is a convex set. (See Solved Exercise 10.) The completeness axiom asserts the converse.

IV. Completeness Axiom Every convex set of real numbers is an interval.

The force of the completeness axiom lies in the fact that intervals have endpoints. Thus, whenever we can prove a set to be convex, the completeness axiom implies the existence of certain real numbers.

Here is some further motivation for the completeness axiom. Suppose that S is a convex set of real numbers which does not extend infinitely in either direction on the number line. Imagine placing the tips of a pair of calipers on two points of S. (See Fig. 4-1.) If x_1 and x_2 are not endpoints of S, we can imagine spreading the calipers to rest on points y_1 and y_2 in S. If y_1 and y_2 are still not endpoints, we can imagine spreading the calipers more and more until no more spreading is possible. The points beyond which the caliper tips cannot spread must be endpoints of S; they may or may not belong to S.

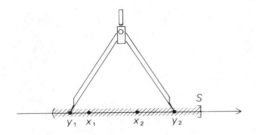

Fig. 4-1 The completeness axiom illustrated by the spreading calipers.

Solved Exercises

10. Prove that $[a, b)$ is convex.

11. Let S be the set consisting of those x for which $0 \leqslant x$ and $x^2 \leqslant 2$. Let c be the right-hand endpoint of S (which exists by the completeness axiom). Prove that $c^2 = 2$; i.e., prove the existence of $\sqrt{2}$. [*Hint*: Show that if $c^2 < 2, 0 < h < 1$, and $0 < h < (2 - c^2)/(2c + 1)$, then $(c + h)^2 < 2$.]

12. Prove that $\sqrt{2}$ is irrational; i.e., show that there is no rational number m/n such that $(m/n)^2 = 2$. [*Hint*: Suppose that m and n have no common factor; is m even or odd?]

13. Prove that any open interval (a, b) contains both rational and irrational numbers.

Exercises

---

20. Let S be the set consisting of those numbers x for which $x \in [0, 1)$ or $x \in (1, 2]$. Prove that S is not convex.

21. Prove that (a, ∞) and $(-\infty, b]$ are convex.

22. For which values of a, b, and c is the set of all x such that $ax^2 + bx + c < 0$ convex? What are the endpoints of the convex set?

23. Prove that, if S and T are convex sets, then the set $S + T$, consisting of all sums $x + y$ with $x \in S$ and $y \in T$, is convex. How are the endpoints of S and T related to the endpoints of $S + T$? If S is open and T is closed, is $S + T$ open or closed?

24. Let A and B be sets of real numbers such that every element of A is less than every element of B, and such that every real number belongs to either A or B. Using the completeness axiom, show that there is exactly one real number c such that every number less than c is in A and every number greater than c is in B.

Problems for Chapter 4 ■■■■■■■■■■■■■■■■■■■■■■■■■■■■■

1. Using the addition and multiplication axioms as stated, prove the following identities.
    (a) $(x + y) + (z + w) = (x + (y + z)) + w$
    (b) $x(a + (b + c)) = xa + (xb + xc)$

2. Prove the following identities.
    (a) $a - (-b) = a + b$
    (b) $(x + y)(u + v) = xu + yu + xv + yv$
    (c) $(ab)^2 = a^2 b^2$
    (d) $\dfrac{1 - x^3}{1 - x} = 1 + x + x^2 \quad (x \neq 1)$

3. Prove the following statements.
    (a) If $x < 0$, then $x^3 < 0$; if $x > 0$, then $x^3 > 0$.
    (b) x and y have the same sign if and only if $xy > 0$.

(c) If $a \leqslant b$, $b \leqslant c$, and $c \leqslant a$, then $a = b = c$.

(d) If $0 < a < b$, then $0 < 1/b < 1/a$.

4. Let $[a_1, b_1]$, $[a_2, b_2]$, $[a_3, b_3]$, ... be an infinite sequence of closed intervals, each of which is contained in the previous one; i.e., $a_i \leqslant a_{i+1}$, $b_{i+1} \leqslant b_i$.

    (a) Using the completeness axiom, prove that there is a real number c which belongs to *all* the intervals in the sequence. [*Hint*: Consider the set consisting of those x such that $x \leqslant b_n$ for all n.]

    (b) Give a condition on the intervals which will insure that there is exactly one real number which belongs to all the intervals.

    (c) Show that the result in (a) is false if the intervals are open.

    (d) Show that the result in (a) is false if we are working with rational numbers rather than with real numbers.

5. Prove that every real number is less than some positive integer. This result is often referred to as the "Archimedian property." [*Hint*: Consider the set S consisting of those x which are less than some positive integer, and show that S cannot have a finite endpoint.]

6. Which of the following sets are convex?

    (a) All x such that $x^3 < 0$.

    (b) All x such that $x^3 < x$.

    (c) All x such that $x^3 < 4$.

    (d) All the areas of polygons inscribed in the circle $x^2 + y^2 = 1$.

    (e) All x such that the decimal expansion of x begins with 2.95.

7. Prove that $(a + (1/a)) \geqslant 2$ if $a > 0$.

8. Use the axioms for addition and multiplication to prove the following:

    (a) $3 - \frac{1}{2} = 4 - \frac{3}{2}$ (b) $\frac{21}{3} = 6 + 1$

    (c) $4 \cdot 2 + 5 \cdot 6 \neq 31$ (d) $2^3 = 8$

    (e) $-(-(-a)) + a = 0$ (f) $-(a + b) = -a - b$

    (g) $\dfrac{a}{b} - \dfrac{c}{b} = \dfrac{a-c}{b}$, $b \neq 0$ (h) $-\dfrac{a}{b} = \dfrac{-a}{b} = \dfrac{a}{-b}$; $b \neq 0$

9. Let S be the set of x such that $x^3 < 10$. Show that S is convex and describe S as an interval. Discuss how this can be used to prove the existence of $\sqrt[3]{10}$.

10. Prove the following inequality using the order axioms:

$$\frac{a^2 + b^2}{2} \geqslant \left(\frac{a+b}{2}\right)^2, \quad \text{where } a < b \text{ and } b < 0.$$

11. Prove that $\sqrt{2} < \sqrt{3}$. [*Hint*: Assume $\sqrt{2} \geqslant \sqrt{3}$ and derive a contradiction.]

12. If $a > 0$, prove that there is a positive integer n such that $0 < 1/n < a$.

13. Suppose that A and B are convex sets such that every element of A is less than every element of B. Show that, if for every positive number ϵ there are

elements a in A and b in B with $b - a < \epsilon$, then there is a transition point from A to B. [*Hint*: Use the completeness axiom to show that A has a right-hand endpoint and B has a left-hand endpoint; then show that these endpoints are equal.]

14. Look up the "least upper bound" version of the completeness axiom* and prove that it is equivalent to ours.

---

*See, for example, Spivak, M. *Calculus*, Publish or Perish, Inc. (1980).

5 Continuity

This chapter defines continuity and develops its basic properties, again without recourse to limits. We shall discuss limits in Chapter 13.

The Definition of Continuity

Naively, we think of a curve as being continuous if we can draw it "without removing the pencil from the paper." Let (x_0, y_0) be a point on the curve, and draw the lines $y = c_1$ and $y = c_2$ with $c_1 < y_0 < c_2$. If the curve is continuous, at least a "piece" of the curve on each side of (x_0, y_0) should be between these lines, as in Fig. 5-1 (left). Compare this with the behavior of the discontinuous curve in Fig. 5-1 (right). The following definition is a precise formulation, for functions, of this idea.

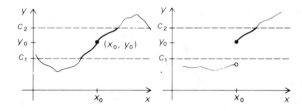

Fig. 5-1 A continuous curve (left) and a discontinuous curve (right).

Definition If x_0 is an element of the domain D of a function f, we say that f is *continuous at x_0* if:

1. For each $c_1 < f(x_0)$ there is an open interval I about x_0 such that, for those x in I *which also lie in D, $c_1 < f(x)$.*

2. For each $c_2 > f(x_0)$ there is an open interval J about x_0 such that, for those x in J *which also lie in D, $f(x) < c_2$.*

If f is continuous at every point of its domain, we simply say that f is continuous or f is *continuous on D.*

Warning It is tempting to define a continuous motion $f(t)$ as one which never passes from $f(t_0)$ to $f(t_1)$ without going through every point between these two. This is a desirable property, but for technical reasons it is not suitable as a definition of continuity; (See Figure 5-4).

The property by which continuity is defined might be called the "principle of persistence of inequalities": f is continuous at x_0 when every strict inequality which is satisfied by $f(x_0)$ continues to be satisfied by $f(x)$ for x in some open interval about x_0. The intervals I and J in the definition may depend upon the value of c_1 and c_2. The definition of continuity may also be phrased in terms of transitions using the idea of Solved Exercise 2, Chapter 13.

Another way to paraphrase the definition of continuity is to say that $f(x)$ is close to $f(x_0)$ when x is close to x_0. The lines $y = c_1$ and $y = c_2$ in Fig. 5-1 provide a measure of closeness. The following example illustrates this idea.

Worked Example 1   The mass y (in grams) of a silver plate which is deposited on a wire during a plating process is given by a function $f(x)$, where x is the length of time (in seconds) during which the plating apparatus is allowed to operate. Suppose that you wish to deposit 2 grams of silver on the wire and that $f(3) = 2$. Being realistic, you know that you cannot control the time *precisely*, but you are willing to accept the result if the mass is less than 0.003 gram in error. Show that if f is continuous, there is a certain tolerance τ such that, if the time is within τ of 3 seconds, the resulting mass of silver plate will be acceptable.

Solution   We wish to restrict x so that $f(x)$ will satisfy the inequalities $1.997 < f(x) < 2.003$. We apply the definition of continuity, with $x_0 = 3$, $c_1 = 1.997$, and $c_2 = 2.003$. From condition 1 of the definition, there is an open interval I containing 3 such that $1.997 < f(x)$ for all $x \in I$. From condition 2, there is J such that $f(x) < 2.003$ for all $x \in J$. For τ less than the distance from 3 to either endpoint of I or J, the interval $[3 - \tau, 3 + \tau]$ is contained in both I and J; for x in this interval, we have, therefore, $1.997 < f(x) < 2.003$.

Of course, to get a specific value of τ which works, we must know more about the function f. Continuity tells us only that such a tolerance τ exists.

Theorem 1, which appears later in this chapter, gives an easy way to verify that many functions are continuous. First, though, we try out the definition on a few simple cases in the following exercises.

Solved Exercises*

1. Let $g(x)$ be the *step function* defined by
$$g(x) = \begin{cases} 0 & \text{if } x \leqslant 0 \\ 1 & \text{if } x > 0 \end{cases}$$
Show that g is not continuous at $x_0 = 0$.

---

*Solutions appear in the Appendix.

2. Let $f(x)$ be the *absolute value function*, $f(x) = |x|$. Show that f is continuous at $x_0 = 0$.

3. Let f be continuous at x_0 and suppose that $f(x_0) \neq 0$. Show that $1/f(x)$ is defined on an open interval about x_0.

4. Decide whether each of the functions whose graphs appear in Fig. 5-2 is continuous. Explain your answers.

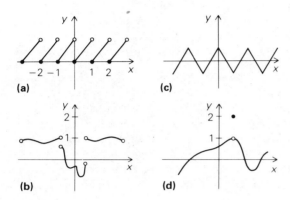

(a)

(c)

(b)

(d)

Fig. 5-2 Which functions are continuous?

Exercises

1. Let $f(x)$ be the step function defined by

$$f(x) = \begin{cases} -1 & \text{if } x < 0 \\ \\ 2 & \text{if } x \geq 0 \end{cases}$$

Show that f is discontinuous at 0.

2. Show that, for any constants a and b, the linear function $f(x) = ax + b$ is continuous at $x_0 = 2$.

3. Let $f(x)$ be defined by

$$f(x) = \begin{cases} x^2 + 1 & \text{if } x < 1 \\ ? & \text{if } 1 \leq x \leq 3 \\ x - 6 & \text{if } 3 < x \end{cases}$$

How can you define $f(x)$ on the interval $[1,3]$ in order to make f continuous on $(-\infty, \infty)$? (A geometric argument will suffice.)

4. Let $f(x)$ be defined by $f(x) = (x^2 - 1)/(x - 1)$ for $x \neq 1$. How should you define $f(1)$ to make the resulting function continuous? [*Hint*: Plot a graph of $f(x)$ for x near 1 by factoring the numerator.]

5. Let $f(x)$ be defined by $f(x) = 1/x$ for $x \neq 0$. Is there any way to define $f(0)$ so that the resulting function will be continuous?

6. Prove from the definition that the function $s(x) = x^2 + 1$ is continuous at 0.

Differentiability and Continuity

If a function $f(x)$ is differentiable at $x = x_0$, then the graph of f has a tangent line at $(x_0, f(x_0))$. Our intuition suggests that if a curve is smooth enough to have a tangent line then the curve should have no breaks—that is, a differentiable function is continuous. The following theorem says just that.

Theorem 1 *If the function f is differentiable at x_0, then f is continuous at x_0.*

Proof We need to verify that conditions 1 and 2 of the definition of continuity hold, under the assumption that the definition of differentiability is met.

We begin by verifying condition 2, so let c_2 be any number such that $f(x_0) < c_2$. We shall produce an open interval I about x_0 such that $f(x) < c_2$ for all x in I.

Choose a positive number M such that $-M < f'(x_0) < M$, and let l_- and l_+ be the lines through $(x_0, f(x_0))$ with slopes $-M$ and M. Referring to Fig. 5-3, we see that l_+ lies below the horizontal line $y = c_2$ for a certain

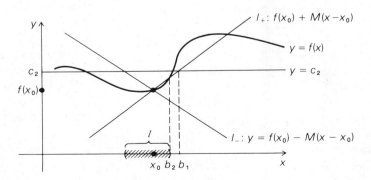

Fig. 5-3 The geometry needed for the proof of Theorem 1.

distance to the right of x_0, and that the graph of f lies below l_+ for a certain distance to the right of x_0 because l_+ overtakes the graph of f at x_0. More precisely, the line $l_+: y = f(x_0) + M(x - x_0)$ intersects $y = c_2$ at

$$b_1 = \frac{c_2 - f(x_0)}{M} + x_0 > x_0$$

and $f(x_0) + M(x - x_0) < c_2$ if $x < b_1$. (The reader should verify this.) Let (a_2, b_2) be an interval which works for l_+ overtaking the graph of f at x_0, so that $f(x) < f(x_0) + M(x - x_0)$ for $x \in (x_0, b_2)$.

If b is the smaller of b_1 and b_2, then

$$f(x) < f(x_0) + M(x - x_0) < c_2 \quad \text{for } x_0 < x < b \tag{1}$$

Similarly, by using the line l_- to the left of x_0, we may find $a < x_0$ such that

$$f(x) < f(x_0) - M(x - x_0) < c_2 \quad \text{for } a < x < x_0 \tag{2}$$

(The reader may wish to add the appropriate lines to Fig. 5-3.) Let $I = (a, b)$. Then inequalities (1) and (2), together with the assumption $f(x_0) < c_2$, imply that

$$f(x) < c_2 \quad \text{for } x \in I,$$

so condition 2 of the definition of continuity is verified.

Condition 1 is verified in an analogous manner. One begins with $c_1 < f(x_0)$ and uses the line l_+ to the left of x_0 and l_- to the right of x_0. We leave the details to the reader.

Worked Example 2 Show that the function $f(x) = (x - 1)/3x^2$ is continuous at $x_0 = 4$.

Solution We know from Chapter 3 that x, $x - 1$, x^2, $3x^2$, and hence $(x - 1)/3x^2$ are differentiable (when $x \neq 0$). Since $4 \neq 0$, Theorem 1 implies that f is continuous at 4.

This method is certainly much easier than attempting to verify directly the conditions in the definition of continuity.

The argument used in this example leads to the following general result.

Corollary

1. *Any polynomial P(x) is continuous.*

2. *Let P(x) and Q(x) be polynomials, with Q(x) not identically zero. Then the rational function R(x) = P(x)/Q(x) is continuous at all points of its domain; i.e., R is continuous at all x_0 such that $Q(x_0) \neq 0$.*

In Chapter 3, we proved theorems concerning sums, products, and quotients of differentiable functions. One can do the same for continuous functions: the sum product and quotient (where the denominator is nonzero) of continuous functions is continuous. Using such theorems one can proceed directly, without recourse to differentiability, to prove that polynomials are continuous. Interested readers can try to work these theorems out for themselves (see Exercise 11 below) or else wait until Chapter 13, where they will be discussed in connection with the theory of limits.

Solved Exercises

5. Prove that $(x^2 - 1)/(x^3 + 3x)$ is continuous at $x = 1$.

6. Is the converse of Theorem 1 true; i.e., is a function which is continuous at x_0 necessarily differentiable there? Prove or give an example.

7. Prove that there is a number $\delta > 0$ such that $x^3 + 8x^2 + x < 1/1000$ if $0 \leqslant x < \delta$.

8. Let f be continuous at x_0 and A a constant. Prove that $f(x) + A$ is continuous at x_0.

Exercises

7. Why can't we ask whether the function $(x^3 - 1)/(x^2 - 1)$ is continuous at 1?

8. Let $f(x) = \dfrac{1}{x} + \dfrac{x^2 - 1}{x}$.

Can you define $f(0)$ so that the resulting function is continuous at all x?

9. Find a function which is continuous on the whole real line, and which is differentiable for all x except 1, 2, and 3. (A sketch will do.)

10. In Solved Exercise 7, show that $\delta = 1/2000$ works; i.e., $x^3 + 8x^2 + x < 1/1000$ if $0 \leqslant x < 1/2000$.

11. (a) Prove that if $f(x) < c_1$ for all x in I and $g(x) < c_2$ for all x in I, then $(f + g)(x) < c_1 + c_2$ for all x in I.

    (b) Prove that, if f and g are continuous at x_0, so is $f + g$.

12.

Let $f(x) = \begin{cases} x^2 & \text{if } x \text{ is rational} \\ 0 & \text{if } x \text{ is irrational} \end{cases}$

    (a) At which points is f continuous?

    (b) At which points is f differentiable?

13. Let f be defined in an open interval about x_0. Suppose that $f(x) = f(x_0) + (x - x_0)g(x)$, where g is continuous at a. Prove that f is differentiable at x_0 and that $f'(x_0) = g(x_0)$. [*Hint*: Prove that $(x - x_0)(g(x) - g(a))$ vanishes rapidly at x_0.]

The Intermediate Value Theorem

A function f is said to have the *intermediate value property* if, whenever f is defined on $[x_1, x_2]$, then $f(x)$ takes every value between $f(x_1)$ and $f(x_2)$ as x runs from x_1 to x_2. Our intuitive notions of continuity suggest that every continuous function has the intermediate value property, and indeed we will prove that this is true. Unfortunately, the intermediate value property is not suitable as a *definition* of continuity; in Fig. 5-4 we have sketched the graph of a function which has the intermediate value property but which is not continuous at 0.

Before, proving the main intermediate value theorem, it is convenient to begin by proving an alternative version.

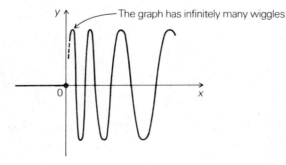

The graph has infinitely many wiggles

Fig. 5-4 A discontinuous function which has the intermediate value property.

Lemma *Intermediate Value Theorem (alternative version). Suppose that f is continuous on $[a, b]$ and that $f(a)$ is less [greater] than some number d. If $f(x) \neq d$ for all $x \in [a, b]$, then $f(b)$ is less [greater] than d as well.*

Proof We write out the proof for the case $f(a) < d$, leaving the case $f(a) > d$ to the reader. We look at the set S consisting of all those x in $[a, b]$ for which $f < d$ on $[a, x]$; i.e., $x \in S$ means that $x \in [a, b]$ and that $f(z) < d$ for all z in $[a, x]$. The idea of the proof is to show that S is an interval and then to prove that $S = [a, b]$. The completeness axiom states that every convex set is an interval. Thus, we begin by proving that S is convex; i.e., if $x_1 < y < x_2$, and x_1 and x_2 are in S, then y is in S as well. First of all, $x_1 \in S$ implies $a \leq x_1$, and $x_2 \in S$ implies $x_2 \leq b$, so we have $a \leq x_1 < y < x_2 \leq b$, so $y \in [a, b]$. To prove that $y \in S$, then, we must show that $f < d$ on $[a, y]$. But $f < d$ on $[a, x_2]$, since $x_2 \in S$, and $[a, y]$ is contained in $[a, x_2]$; thus, for $z \in [a, y]$, we have $z \in [a, x_2]$, so $f(z) < d$; i.e., $f < d$ on $[a, y]$. We have proven that $y \in S$, so S is convex.

By the completeness axiom, S is an interval. Since $f(a) < d$, we have $f < d$ on $[a, a]$, so $a \in S$. Nothing less than a can be in S, so a is the left-hand endpoint of S. Since S is contained in $[a, b]$, it cannot extend infinitely far to the right; we conclude that $S = [a, c)$ or $[a, c]$ for some c in $[a, b]$.

Case 1 Suppose $S = [a, c)$. Then, for every $z \in [a, c)$, we have $f < d$ on $[a, z]$, so $f(z) < d$; we have thus shown that $f < d$ on $[a, c)$. If $f(c)$ were less than d, we would have $f < d$ on $[a, c]$, so that c would belong to S, contradicting the statement that $S = [a, c)$. $f(c)$ cannot equal d, by the assumption of the theorem we are proving, so the only remaining possibility is $f(c) > d$. By the continuity of f at c, we must have $f(x) > d$ for all x in some open interval about c; but this contradicts the fact that $f < d$ on $[a, c)$. We conclude that the case $S = [a, c)$ simply cannot occur.

Case 2 Suppose that $S = [a, c]$, with $c < b$. Since $c \in S$, we have $f < d$ on $[a, c]$. By the continuity of f at c and the fact that $f(c) < d$, we conclude that $f < d$ also on some open interval (p, q) containing c. (Here we use the fact that $c < b$, so that f is defined on an open interval about c.) But if $f(z) < d$ for z in $[a, c]$ and for z in (p, q), with $p < c < q \leq b$, then $f < d$ on $[a, q)$. Let $y = \frac{1}{2}(c + q)$, the point halfway between c and q. Then $f < d$ on $[a, y]$, so $y \in S$. Since $y > c$, this contradicts the statement that $S = [a, c]$. Thus, case 2 cannot occur.

The only possibility which remains, and which therefore must be true, is that $S = [a, b]$. Thus, $f(z) < d$ for all $z \in [a, b]$, and in particular for $z = b$, so $f(b) < d$, which is what was to be proven.

We can now deduce the usual form of the intermediate value theorem.

Theorem 2 *Intermediate Value Theorem* *Let f be continuous on* $[a, b]$, *and assume that* $f(a) < d$ *and* $f(b) > d$, *or* $f(a) > d$ *and* $f(b) < d$. *Then there exists a number c in* (a, b) *such that* $f(c) = d$.

Proof If there were no such c, then the alternative version, which we have just proven, would enable us to conclude that $f(a)$ and $f(b)$ lie on the same side of d. Since we have assumed in the statement of the theorem that $f(a)$ and $f(b)$ lie on opposite sides of d, the absence of a c such that $f(c) = d$ leads to a contradiction, so the c must exist.

Solved Exercises

9. In the proof of the intermediate value theorem, why did we not use, instead of S, the set T consisting of those x in $[a, b]$ for which $f(x) < d$?

10. Find a formula for a function like that shown in Fig. 5-4. (You may use trigonometric functions.)

11. Prove that the polynomial $x^5 + x^4 - 3x^2 + 2x + 8$ has at least one real root.

12. Let T be the set of *values* of a function f which is continuous on $[a, b]$; i.e., $y \in T$ if and only if $y = f(x)$ for some $x \in [a, b]$. Prove that T is convex.

Exercises

14. Let f be a polynomial and suppose that $f'(-1) < 0$ while $f'(1) > 0$. Prove that f must have a critical point (a point where f' vanishes) somewhere on the interval $(-1, 1)$.

15. Let $f(x) = x^4 - x^2 + 35x - 7$. Prove that f has at least two real roots.

16. (a) Give a direct proof of the Intermediate Value Theorem.

    (b) Use the intermediate value theorem to prove the alternative version.

Increasing and Decreasing at a Point

Intuitively, if we say $f(x)$ is increasing with x at x_0, we mean that when x is increased a little, $f(x)$ increases and when x is decreased a little, $f(x)$ decreases. Study Fig. 5-5 to see why we say "a little."

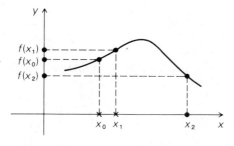

Fig. 5-5 x_1 is a little greater than x_0 and $f(x_1)$ $> f(x_0)$; x_2 is a lot greater than x_0 and now $f(x_2) <$ $f(x_0)$.

Definition Let f be a function whose domain contains an open interval about x_0.

    1. f is said to be *increasing* at x_0 if the graph of f overtakes the horizontal line through $(x_0, f(x_0))$ at x_0.

    2. f is said to be *decreasing* at x_0 if the graph is overtaken by the line at x_0. (See Fig. 5-6.)

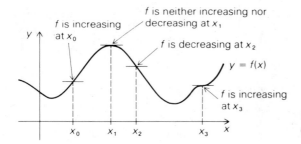

Fig. 5-6 Where is f increasing and decreasing?

    Thus, f is increasing at x_0 if and only if $f(x) - f(x_0)$ changes sign from negative to positive at x_0. Similarly f is decreasing if and only if $f(x) - f(x_0)$ changes sign from positive to negative at x_0.

    If we substitute the definition of "overtake" in the definition of increasing, we obtain the following equivalent reformulation.

Definition' Let f be a function whose domain contains an open interval about x_0.

    1. f is said to be *increasing* at x_0 if there is an open interval I about x_0 such that:

(i) $f(x) < f(x_0)$ for $x < x_0$ in I

(ii) $f(x) > f(x_0)$ for $x > x_0$ in I

2. f is said to be *decreasing* at x_0 if there is an open interval I about x_0 such that:

(i) $f(x) > f(x_0)$ for $x < x_0$ in I

(ii) $f(x) < f(x_0)$ for $x > x_0$ in I

Pictorially speaking, a function f is increasing at x_0 when moving x a little to the left of x_0 lowers $f(x)$ while moving x a little to the right of x_0 raises $f(x)$. (The opposite happens if the function is decreasing at x_0.)

Worked Example 3 Show that $f(x) = x^2$ is increasing at $x_0 = 2$.

Solution Let I be $(1, 3)$. If $x < x_0$ is in I, we have $1 < x < 2$, so $f(x) = x^2 < 4 = x_0^2$. If $x > x_0$ is in I, then $2 < x < 3$, and so $f(x) = x^2 > 4 = x_0^2$. We have verified (i) and (ii) of part 1 of Definition', so f is increasing at 2.

The transition definition of the derivative of f at x_0 tells us which lines overtake and are overtaken by the graph of f at x_0. This leads to the next theorem.

Theorem 3 *Let f be differentiable at x_0.*

1. If $f'(x_0) > 0$, then f is increasing at x_0.

2. If $f'(x_0) < 0$, then f is decreasing at x_0.

3. If $f'(x_0) = 0$, then f may be increasing at x_0, decreasing at x_0, or neither.

Proof We shall prove parts 1 and 3; the proof of part 2 is similar to 1.

The definition of the derivative, as formulated in Theorem 4, Chapter 2, includes the statement that any line through $(x_0, f(x_0))$ whose slope is less than $f'(x_0)$ is overtaken by the graph of f at x_0. If $f'(x_0) > 0$, then the horizontal line through $(x_0, f(x_0))$, whose slope is 0, must be overtaken by the graph of f at x_0; thus, f is increasing at x_0, by definition.

The functions x^3, $-x^3$, and x^2, all of which have derivative 0 at $x_0 = 0$, establish part 3; see Solved Exercise 15.

The following is another proof of Theorem 3 directly using the original definition of the derivative in Chapter 1.

Alternative Proof From condition 1 of the definition of the derivative (p. 6), we know that if $m < f'(x_0)$, then the function $f(x) - [f(x_0) + m(x - x_0)]$ changes sign from negative to positive at x_0. Thus if $0 < f'(x_0)$, then we may choose $m = 0$ and conclude that $f(x) - f(x_0)$ changes sign from negative to positive at x_0; that is, f is increasing at x_0. This establishes part 1 of the theorem. Part 2 is similar, and the functions x^3, $-x^3$, and x^2 establish part 3 as in the first proof.

Worked Example 4 Is $x^5 - x^3 - 2x^2$ increasing or decreasing at -2?

Solution Letting $f(x) = x^5 - x^3 - 2x^2$, we have $f'(x) = 5x^4 - 3x^2 - 4x$ and $f'(-2) = 5(-2)^4 - 3(-2)^2 - 4(-2) = 80 - 12 + 8 = 76$, which is positive. By Theorem 3 part 3, $x^5 - x^3 - 2x^2$ is increasing at -2.

Theorem 3 can be interpreted geometrically: if the linear approximation to f at x_0 (that is, the tangent line) is an increasing or decreasing function, then f itself is increasing or decreasing at x_0. If the tangent line is horizontal, the behavior of f at x_0 is not determined by the tangent line. (See Fig. 5-7.)

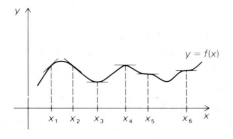

Fig. 5-7
$f'(x_1) > 0$; f is increasing at x_1
$f'(x_2) < 0$; f is decreasing at x_2
$f'(x_3) = f'(x_4) = 0$; f is neither
increasing nor decreasing at x_3 and x_4
$f'(x_5) = 0$; f is decreasing at x_5
$f'(x_6) = 0$; f is increasing at x_6

We can also interpret Theorem 3 in terms of velocities. If $f(t)$ is the position of a particle on the real-number line at time t, and $f'(t_0) > 0$, then the particle is moving to the right at time t_0; if $f'(t_0) < 0$, the particle is moving to the left.

Combined with the techniques for differentiation in Chapter 3, Theorem 3 provides an effective means for deciding where a function is increasing or decreasing.

Solved Exercises

13. The temperature at time x is given by $f(x) = (x + 1)/(x - 1)$ for $x \neq 1$. Is it getting warmer or colder at $x = 0$?

14. Using Theorem 3, find the points at which $f(x) = 2x^3 - 9x^2 + 12x + 5$ is increasing or decreasing.

15. Decide whether each of the functions x^3, $-x^3$, and x^2 is increasing, decreasing, or neither at $x = 0$.

Exercises

17. If $f(t) = t^5 - t^4 + 2t^2$ is the position of a particle on the real-number line at time t, is it moving to the left or right at $t = 1$?

18. Find the points at which $f(x) = x^2 - 1$ is increasing or decreasing.

19. Find the points at which $x^3 - 3x^2 + 2x$ is increasing or decreasing.

20. Is $f(x) = 1/(x^2 + 1)$ increasing or decreasing at $x = 1, -3, \frac{3}{4}, 25, -36$?

21. A ball is thrown upward with an initial velocity of 30 meters per second. The ball's height above the ground at time t is $h(t) = 30t - 4.9t^2$. When is the ball rising? When is it falling?

22. (a) Prove that, if $f(x_0) = 0$, then f is increasing [decreasing] at x_0 if and only if $f(x)$ changes sign from negative to positive [positive to negative] at x_0.

    (b) Prove that f is increasing [decreasing] at x_0 if and only if $f(x) - f(x_0)$ changes sign from negative to positive [positive to negative] at x_0.

Increasing or Decreasing on an Interval

Suppose that f is increasing at every point of an interval $[a, b]$. We would expect $f(b)$ to be larger than $f(a)$. In fact, we have the following useful result.

Theorem 4   *Let f be continuous on $[a, b]$, where $a < b$, and suppose that f is increasing [decreasing] at all points of (a, b). Then $f(b) > f(a)$ $[f(b) < f(a)]$.*

The statement of Theorem 4 may appear to be tautological—that is, "trivially true"—but in fact it requires a proof (which will be given shortly). Like the intermediate value theorem, Theorem 4 connects a *local* property of functions (increasing at each point of an interval) with a *global* property (relation between values of the function at endpoints). We do not insist that f be increasing or decreasing at a or b because we wish the theorem to apply in cases of the type illustrated in Fig. 5-8. Also, we note that if f is not continuous, the result is not valid (see Fig. 5-9).

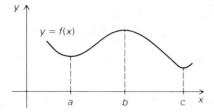

Fig. 5-8 f is increasing at each point of (a,b); $f(b) > f(a)$
f is decreasing at each point of (b,c); $f(c) < f(b)$
f is neither increasing nor decreasing at a, b, c

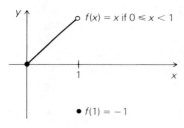

Fig. 5-9 f is increasing at all points of $(0,1)$, but $f(1)$ is not greater than $f(0)$.

Proof of Theorem 4 We proceed in several steps.

Step 1 If $a < x < y < b$, then $f(x) < f(y)$.

To prove this, we choose any x in (a,b) and let S be the set consisting of those y in (x,b) for which $f(x) < f(z)$, for all z in $(x,y]$. If we can show that $S = (x,b)$, then for any y such that $x < y < b$ we will have $y \in S$, so that $f(x) < f(z)$ for all z in $(x,y]$; in particular, we will have $f(x) < f(y)$.

We proceed to show that $S = (x,b)$. By the same kind of argument as we used in the proof of the intermediate value theorem (alternative version) it is easy to show that S is convex. By the completeness axiom, S is an interval. Since f is increasing at x, S contains all the points sufficiently near to x and to the right of x, so x is the left-hand endpoint of S. Thus $S = (x,c)$ or $(x,c]$ for some $c, c \leqslant b$.

Suppose that $c < b$. Then f is increasing at c, so we can find points p and q such that:

$$x < p < c < q < b \tag{1}$$

$$f(y) < f(c) \qquad \text{for all } y \text{ in } (p,c) \tag{2}$$

and

$$f(c) < f(y) \qquad \text{for all } y \text{ in } (c,q) \tag{3}$$

Since $S = (x, c)$, we have $f(x) < f(y)$ for all y in (p, c). By (2), we have $f(x) < f(c)$; by (3), we then have $f(x) < f(y)$ for all y in (c, q). Thus, we have $f(x) < f(y)$ for all y in (x, c) and $[c, q)$, so $f(x) < f$ on (x, q), i.e., $f(x) < f(y)$ for all $y \in (x, q)$. Thus S contains points to the right of c, contradicting the fact that c is the righthand endpoint of S. Hence c must equal b, and so $S = (x, b)$.

Notice that we have not yet used the continuity of f at a and b. (See the corollary below.)

Step 2 If $y \in (a, b)$, then $f(a) \leqslant f(y)$ and $f(y) \leqslant f(b)$.

To prove that $f(a) \leqslant f(y)$, we assume the opposite, namely, $f(a) > f(y)$, and derive a contradiction. In fact, if $f(a) > f(y)$ for some $y \in (a, b)$, the continuity of f at a implies that $f > f(y)$ on some interval $[a, r]$. Choosing x in (a, r) such that $x < y$, we have $f(x) > f(y)$, contradicting step 1. Using the continuity of f at b, we can prove in a similar manner that $f(y) \leqslant f(b)$.

Step 3 If $y \in (a, b)$, then $f(a) < f(y)$ and $f(y) < f(b)$.

To prove that $f(a) < f(y)$, we choose any x between a and y (i.e., let $x = \frac{1}{2}(a + y)$). By step 1, $f(x) < f(y)$; by step 2, $f(a) \leqslant f(x)$. So $f(a) \leqslant f(x) < f(y)$, and $f(a) < f(y)$. Similarly, we prove $f(y) < f(b)$.

Step 4 $f(a) < f(b)$.

Choose any y in (a, b). By step 3, $f(a) < f(y)$ and $f(y) < f(b)$, so $f(a) < f(b)$.

We shall now rephrase Theorem 4. The following terminology will be convenient.

Definition Let f be a function defined on an interval I. If $f(x_1) < f(x_2)$ for all $x_1 < x_2$ in I, we say that f is *increasing* on I. If $f(x_1) > f(x_2)$ for all $x_1 < x_2$ in I, we say that f is *decreasing* on I. If f is either increasing on I or decreasing on I, we say that f is *monotonic* on I.

Theorem 4′ *Let f be continuous on $[a, b]$ and increasing [decreasing] at all points of (a, b). Then f is increasing [decreasing] on $[a, b]$.*

For example, the function in Fig. 5-8 is monotonic on $[a, b]$ and monotonic on $[b, c]$, but it is not monotonic on $[a, c]$. The function $f(x) = x^2$ is monotonic on $(-\infty, 0)$ and $(0, \infty)$, but not on $(-\infty, \infty)$. (Draw a sketch to convince yourself.)

Combining Theorems 3 and 4′ with the intermediate value theorem gives a result which is useful for graphing.

Theorem 5 *Suppose that:*

1. *f is continuous on $[a, b]$.*

2. *f is differentiable on (a, b) and f' is continuous on (a, b).*

3. *f' is never zero on (a, b).*

Then f is monotonic on $[a, b]$. To check whether f is increasing or decreasing on $[a, b]$, it suffices to compute the value of f' at any one point of (a, b) and see whether it is positive or negative.

Proof* By the intermediate value theorem, f' must either be positive everywhere on (a, b) or negative everywhere on (a, b). (If f' took values with both signs, it would have to be zero somewhere in between.) If f' is positive everywhere, f is increasing at each point of (a, b) by Theorem 3. By Theorem 4', f is increasing on $[a, b]$. If f' is negative everywhere, f is decreasing on $[a, b]$.

We can also apply Theorem 5 to intervals which are not closed.

Corollary *Suppose that:*

1. *f is continuous on an open interval (a, b).*

2. *f is differentiable and f' is continuous at each point of (a, b).*

3. *f' is not zero at any point of (a, b).*

Then f is monotonic on (a, b) (increasing if $f' > 0$, decreasing if $f' < 0$).

Proof By the intermediate value theorem, f' is everywhere positive or everywhere negative on (a, b). Suppose it to be everywhere positive. Let $x_1 < x_2$ be in (a, b). We may apply Theorems 3 and 4 to f on $[x_1, x_2]$ and conclude that $f(x_1) < f(x_2)$. Thus f is increasing on (a, b). Similarly, if f' is everywhere negative on (a, b), f is decreasing on (a, b).

Similar statements hold for half-open intervals $[a, b)$ or $(a, b]$.

Applications of these results to the shape of graphs will be given in the next chapter.

---

*Theorem 5 is still true if f' is not continuous; see Problem 10 of Chapter 7.

Solved Exercises

16. The function $f(x) = -2/x$ has $f'(x) = 2/x^2 > 0$ at all points of its domain, but $f(1) = -2$ is not greater than $f(-1) = 2$. What is wrong here?

17. Let $a < b < c$, and suppose that f, defined on (a, c), is increasing at each point of (a, b) and (b, c) and is continuous at b. Prove that f is increasing at b.

Exercises

23. Prove the analogue of Theorem 4 for decreasing functions.

24. Show by example that the conclusion in Solved Exercise 17 may not be valid if f is discontinuous at b.

25. (a) Suppose that the functions f and g defined on $[a, b]$ are continuous at a and b, f' and g' are continuous on (a, b) and that $f'(x) > g'(x)$ for all x in (a, b). If $f(a) = g(a)$, prove that $f(b) > g(b)$. [*Hint*: You may assume that $f - g$ is continuous at a and b.]

    (b) Give a physical interpretation of the result of part (a), letting x be *time*.

26. Let f be a polynomial such that $f(0) = f(1)$ and $f'(0) > 0$. Prove that $f'(x) = 0$ for some x in $(0, 1)$.

The Extreme Value Theorem

The last theorem in this chapter asserts that a continuous function on a closed interval has maximum and minimum values. Again, the proof uses the completeness axiom.

We begin with a lemma which gets us part way to the theorem.

Boundedness Lemma *Let f be continuous on $[a, b]$. Then there is a number B such that $f(x) \leqslant B$ for all x in $[a, b]$. We say that f is bounded above by B on $[a, b]$ (see Fig. 5-10).*

Proof If P is any subset of $[a, b]$, we will say that f is *bounded above* on P if there is some number m (which may depend on P) such that $f(x) \leqslant m$ for all x in P. Let S be the set consisting of those y in $[a, b]$ such that f is bounded above on $[a, y]$. If $y_1 < y < y_2$, where y_1 and y_2 are in S, then f

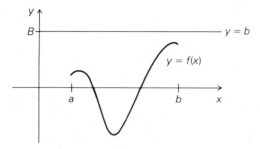

Fig. 5-10 The line $y = B$ lies above the graph of f on $[a, b]$.

is bounded above on $[a, y_2]$, so it must be bounded above on $[a, y]$. Also, y must belong to $[a, b]$ if y_1 and y_2 do, so S is convex.

By the completeness axiom, S is an interval which is contained in $[a, b]$. f is obviously bounded above on $[a, a]$ (let $m = f(a)$), so $a \in S$, and S is of the form $[a, c)$ or $[a, c]$ for some c in $[a, b]$.

Let c be the right-hand endpoint of S. Since f is continuous at e, f is bounded above on some interval about c; let e be the left-hand endpoint of that interval. Since $e \in [a, c)$, f is bounded above on $[a, e]$ as well as on $[e, c]$, so f is bounded above on $[a, c]$, and $c \in S$. Thus, S is of the form $[a, c]$. If $c < b$, an argument like the one we just used shows that S must contain points to the right of c, which is a contradiction. So we must have $c = b$, i.e., $S = [a, b]$, and f is bounded above on $[a, b]$.

By taking the "best" bound for f on $[a, b]$ (imagine lowering the horizontal line in Fig. 5-10 until it just touches the graph of f), we will obtain the maximum value.

Theorem 6 Extreme Value Theorem *Let f be continuous on $[a, b]$. Then f has both a maximum and minimum value on $[a, b]$; i.e., there are points x_M and x_m in $(a, b]$ such that $f(x_m) \leqslant f(x) \leqslant f(x_M)$ for all x in $[a, b]$.*

Proof We prove that there is a maximum value, leaving the case of a minimum to the reader. Consider the set T of values of f, i.e., $q \in T$ if and only if $q = f(x)$ for some x in $[a, b]$. We saw in Solved Exercise 12 that T is convex. By the completeness axiom, T is an interval. By the boundedness lemma, T cannot extend infinitely far in the positive direction, so it has an upper endpoint, which we call M. (See Fig. 5-11.) We wish to show that the graph of f actually touches the line $y = M$ at some point, so that M will be a value of $f(x)$ for some x in $[a, b]$, and thus M will be the maximum value.

If there is no x in $[a, b]$ for which $f(x) = M$, then $f(x) < M$ for all x in $[a, b]$, and an argument identical to that used in the proof of the

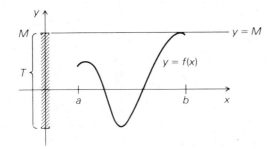

Fig. 5-11 $y = M$ is the maximum value of f on $[a, b]$.

boundedness lemma shows that there is some $M_1 < M$ such that $f(x) \leqslant M_1$ for all x in $[a, b]$. But this contradicts the assumption that M is the upper endpoint of the set T of values of f, so there must be some x_M in $[a, b]$ for which $f(x_M) = M$. For any y in $[a, b]$, we now have $f(y) \in T$, so $f(y) \leqslant M = f(x_M)$, and so M is the maximum value of f on $[a, b]$.

Solved Exercises

18. Prove that there exists a number M such that $x^8 + x^4 + 8x^9 - x < M$ if $0 \leqslant x \leqslant 10,000$.

19. Prove that, if f is continuous on $[a, b]$, then the set T of values of f (see Solved Exercise 12) is a closed interval. Must $f(a)$ and $f(b)$ be the endpoints of this interval?

Exercises

27. (a) Prove. that, if f is continuous on $[a, b]$, so is the function $-f$ defined by $(-f)(x) = -[f(x)]$.

   (b) Prove the "minimum" part of the extreme value theorem by applying the "maximum" part to $-f$.

28. Find a specific number M which works in Solved Exercise 18.

29. Give an example of a continuous function f on $[0, 1]$ such that neither $f(0)$ nor $f(1)$ is an endpoint of the set of values of f on $[0, 1]$.

30. Is the boundedness lemma true if the closed interval $[a, b]$ is replaced by an open interval (a, b)?

Problems for Chapter 5

1. We can define all the notions of this section, including continuity, differentiability, maximum and minimum values, etc., for functions of a *rational* variable; i.e., we may replace real numbers by rationals everywhere in the definitions. In particular, $[a, b]$ then means the set of rational x for which $a \leqslant x \leqslant b$.

   (a) Prove that the function

   $$f(x) = \frac{1}{x^2 - 2}$$

   is defined everywhere on the rational interval $[0, 2]$. It is possible to prove that f is continuous on $[0, 2]$; you may assume this now.

   (b) Show that the function f in part (a) does not satisfy the conclusions of the intermediate value theorem, the boundedness lemma, or the extreme value theorem. Thus, it is really necessary to work with the real numbers.

2. Prove that, if f is a continuous function on an interval I (not necessarily closed), then the set of values of f on I is an interval. Could the set of values be a closed interval even if I is not?

3. Prove that the volume of a cube is a continuous function of the length of its edges.

4. "Prove" that you were once exactly 1 meter tall. Did you ever weigh 15 kilograms?

5. Write a direct proof of the "minimum" part of the extreme value theorem.

6. Prove that, if f and g are continuous on $[a, b]$, then so is $f - g$.

7. Assuming the result of Problem 6, prove that, if $f(a) < a$, $f(b) > b$, and f is continuous on $[a, b]$, then there is some x in $[a, b]$ such that $f(x) = x$.

8. Give an example of (discontinuous) functions f and g on $[0, 1]$ such that f and g both have maximum values on $[0, 1]$, but $f + g$ does not.

9. Let $f(x) = x^{17} + 3x^4 - 2$ and $g(x) = 5x^6 - 10x + 3$. Prove that there is a number x_0 such that $f(x_0) = g(x_0)$.

10. (a) Prove that, if f is increasing at each point of an open interval I, then the set S consisting of all those x in I for which $f(x) \in (0, 1)$ is convex.

    (b) Show that S might not be convex if f is not increasing on I.

11. Show by example that the sum of two functions with the intermediate value property need not have this property.

12. Prove that the intermediate value theorem implies the completeness axiom. [*Hint*: If there were a convex set which was not an interval, show that you could construct a continuous function which takes on only the values 0 and 1.]

6 Graphing

In this chapter we continue our development of differential calculus by analyzing the shape of graphs. For extensive examples, consult your regular calculus text. We will give an example not usually found in calculus texts: how to graph a *general* cubic or quartic.

Turning Points

Points where the derivative f' changes sign are called *turning points* of f; they separate the intervals on which f is increasing and decreasing.

> **Definition** If f is differentiable at x_0 and $f'(x_0) = 0$, we call x_0 a *critical point* of f.
>
> If f is defined and differentiable throughout an open interval containing x_0, we call x_0 a *turning point* of f if f' changes sign at x_0.

If f' is continuous at x_0, then a turning point is also a critical point (why?). A critical point *need not* be a turning point, however, as the function $y = x^3$ shows.

To investigate the behavior of the graph of f at a turning point, it is useful to treat separately the two possible kinds of sign change. Suppose first that f' changes sign from negative to positive at x_0. Then there is an open interval (a, b) about x_0 such that $f'(x) < 0$ on (a, x_0) and $f'(x) > 0$ on (x_0, b). Applying the remark following the corollary to Theorem 5 of Chapter 5, we find that f is decreasing on $(a, x_0]$ and increasing on $[x_0, b)$. It follows that $f(x) > f(x_0)$ for x in (a, x_0) and $f(x) > f(x_0)$ for x in (x_0, b). (See Fig. 6-1.)

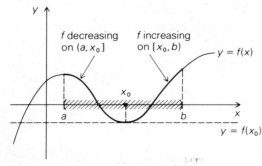

Fig. 6-1 f has a local minimum point at x_0.

Notice that for all $x \neq x_0$ in (a, b), $f(x_0) < f(x)$; that is, the smallest value taken on by f in (a, b) is achieved at x_0. For this reason, we call x_0 a *local minimum point* for f.

In case f' changes from positive to negative at x_0, a similar argument shows that f behaves as shown in Fig. 6-2 with $f(x_0) > f(x)$ for all $x \neq x_0$ in an open interval (a, b) around x_0. In this case, x_0 is called a *local maximum point* for f.

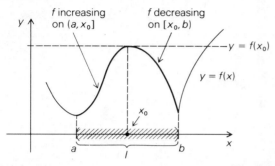

Fig. 6-2 f has a local maximum point at x_0.

Worked Example 1 Find and classify the turning points of the function $f(x) = x^3 + 3x^2 - 6x$.

Solution The derivative $f'(x) = 3x^2 + 6x - 6$ has roots at $-1 \pm \sqrt{3}$; it is positive on $(-\infty, -1 - \sqrt{3})$ and $(-1 + \sqrt{3}, \infty)$ and is negative on $(-1 - \sqrt{3}, -1 + \sqrt{3})$. Changes of sign occur at $-1 - \sqrt{3}$ (positive to negative) and $-1 + \sqrt{3}$ (negative to positive), so $-1 - \sqrt{3}$ is a local maximum point and $-1 + \sqrt{3}$ is a local minimum point.

We can summarize the results we have obtained as follows.

Theorem 1 *Suppose that f' is continuous in an open interval containing x_0.*

1. *If the sign change of f' is from negative to positive, then x_0 is a local minimum point.*

2. *If the sign change of f' is from positive to negative, then x_0 is a local maximum point.*

3. *If f' is continuous, then every turning point is a critical point, but a critical point is not necessarily a turning point.*

Solved Exercises

1. Find the turning points of the function $f(x) = 3x^4 - 8x^3 + 6x^2 - 1$. Are they local maximum or minimum points?

2. Find the turning points of x^n for $n = 0, 1, 2, \ldots$.

3. Does $g(x) = 1/f(x)$ have the same turning points as $f(x)$?

Exercises

1. For each of the functions in Fig. 6-3, tell whether x_0 is a turning point, a local minimum point, or a local maximum point.

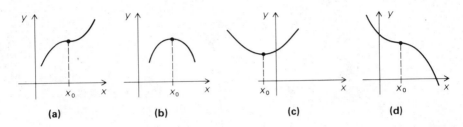

Fig. 6-3 Identify the turning points.

2. Find the turning points of each of the following functions and tell whether they are local maximum or minimum points:

   (a) $f(x) = \frac{1}{3}x^3 - \frac{5}{2}x^2 + 4x + 1$ (b) $f(x) = 1/x^2$

   (c) $f(x) = ax^2 + bx + c$ $(a \neq 0)$ (d) $f(x) = 1/(x^4 - 2x^2 - 5)$

3. Find a function f with turning points at $x = -1$ and $x = \frac{1}{2}$.

4. Can a function be increasing at a turning point? Explain.

A Test for Turning Points

If g is a function such that $g(x_0) = 0$ and $g'(x_0) > 0$, then g is increasing at x_0, so g changes sign from negative to positive at x_0. Similarly, if $g(x_0) = 0$ and $g'(x_0) < 0$, then g changes sign from positive to negative at x_0. Applying these assertions to $g = f'$, we obtain the following useful result.

Theorem 2   Second Derivative Test *Let f be a differentiable function, let x_0 be a critical point of f (that is, $f'(x_0) = 0$), and assume that $f''(x_0)$ exists.*

   1. If $f''(x_0) > 0$, then x_0 is a local minimum point for f.

2. *If $f''(x_0) < 0$, then x_0 is a local maximum point for f.*

3. *If $f''(x_0) = 0$, then x_0 may be a local maximum or minimum point, or it may not be a turning point at all.*

Proof For parts 1 and 2, we use the definition of turning point as a point where the derivative changes sign. For part 3, we observe that the functions x^4, $-x^4$, and x^3 all have $f'(0) = f''(0) = 0$, but zero is a local minimum point for x^4, a local maximum point for $-x^4$, and not a turning point for x^3.

Worked Example 2 Use the second derivative test to analyze the critical points of $f(x) = x^3 - 6x^2 + 10$.

Solution Since $f'(x) = 3x^2 - 12x = 3x(x - 4)$, the critical points are at 0 and 4. Since $f''(x) = 6x - 12$, we find that $f''(0) = -12 < 0$ and $f''(4) = 12 > 0$. By Theorem 2, zero is a local maximum point and 4 is a local minimum point.

When $f''(x_0) = 0$ the second derivative test is inconclusive. We may still use the first derivative test to analyze the critical points, however.

Worked Example 3 Analyze the critical point $x_0 = -1$ of $f(x) = 2x^4 + 8x^3 + 12x^2 + 8x + 7$.

Solution The derivative is $f'(x) = 8x^3 + 24x^2 + 24x + 8$, and $f'(-1) = -8 + 24 - 24 + 8 = 0$, so -1 *is* a critical point. Now $f''(x) = 24x^2 + 48x + 24$, so $f''(-1) = 24 - 48 + 24 = 0$, and the second derivative test is inconclusive. If we factor f', we find $f'(x) = 8(x^3 + 3x^2 + 3x + 1) = 8(x + 1)^3$. Thus -1 is the only root of f', $f'(-2) = -8$, and $f'(0) = 8$, so f' changes sign from negative to positive at -1; hence -1 is a local minimum point for f.

Solved Exercises

4. Analyze the critical points of $f(x) = x^3 - x$.

5. Show that if $f'(x_0) = f''(x_0) = 0$ and $f'''(x_0) > 0$, then x_0 is *not* a turning point of f.

Exercises

5. Analyze the critical points of the following functions:

   (a) $f(x) = x^4 - x^2$

   (b) $g(s) = s/(1 + s^2)$

(c) $h(p) = p + (1/p)$

6. Show that if $f'(x_0) = f''(x_0) = 0$ and $f'''(x_0) < 0$, then x_0 is *not* a turning point for f.

Concavity

The sign of $f''(x_0)$ has a useful interpretation, even if $f'(x_0)$ is not zero, in terms of the way in which the graph of f is "bending" at x_0. We first make a preliminary definition.

Definition If f and g are functions defined on a set D containing x_0, we say that *the graph of f lies above the graph of g near x_0* if there is an open interval I about x_0 such that, for all $x \neq x_0$ which lie in I and D, $f(x) > g(x)$.

The concepts of continuity, local maximum, and local minimum can be expressed in terms of the notion just defined, where one of the graphs is a horizontal line, as follows:

1. f is continuous at x_0 if
   (a) For every $c > f(x_0)$, the graph of f lies below the line $y = c$ near x_0, and
   (b) For every $c < f(x_0)$, the graph of f lies above the line $y = c$ near x_0.

2. x_0 is a local maximum point for f if the graph of f lies below the line $y = f(x_0)$ at x_0.

3. x_0 is a local minimum point for f if the graph of f lies above the line $y = f(x_0)$ at x_0.

Now we use the notion of lying above and below to define upward and downward concavity.

Definition Let f be defined in an open interval containing x_0. We say that f is *concave upward* at x_0 if, near x_0, the graph of f lies above some line through $(x_0, f(x_0))$, and that f is *concave downward* at x_0 if, near x_0, the graph of f lies below some line through $(x_0, f(x_0))$.

Study Fig. 6-4. The curve "holds water" where it is concave upward and "sheds water" where it is concave downward. Notice that the curve may lie

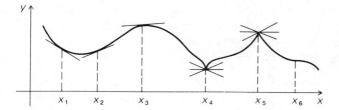

Fig. 6-4 f is concave upward at x_1, x_2, x_4, downward at x_3, x_5, neither at x_6.

above or below several different lines at a given point (like x_4 or x_5); however, if the function is differentiable at x_0, the line must be the tangent line, since the graph either overtakes or is overtaken by every other line.

It appears from Fig. 6-4 that f is concave upward at points where f' is increasing and concave downward at points where f' is decreasing. This is true in general.

Theorem 3 *Suppose that f is differentiable on an open interval about x_0. If $f'(x)$ is increasing at x_0, then f is concave upward at x_0. If $f'(x)$ is decreasing at x_0, then f is concave downward at x_0.*

Proof We reduce the problem to the case where the derivative is zero by looking at the difference between $f(x)$ and its first-order approximation. Let $l(x) = f(x_0) + f'(x_0)(x - x_0)$, and let $r(x) = f(x) - l(x)$. (See Fig. 6-5.) Differentiating, we find $l'(x) = f'(x_0)$, and $r'(x) = f'(x) - l'(x) = f'(x) - f'(x_0)$. In particular, $r'(x_0) = f'(x_0) - f'(x_0) = 0$.

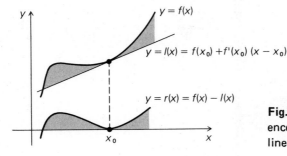

Fig. 6-5 $r(x)$ is the difference between $f(x)$ and its linear approximation.

Suppose that f' is increasing at x_0. Then r', which differs from f' by the constant $f'(x_0)$, is also increasing at x_0. Since $r'(x_0) = 0$, r' must change sign from negative to positive at x_0. Thus, x_0 is a turning point for r; in fact, it is a local minimum point. This means that, for $x \neq x_0$ in some open interval I about x_0, $r(x) > r(x_0)$. But $r(x_0) = f(x_0) - l(x_0) =$

$f(x_0) - f(x_0) = 0$, so $r(x) > 0$ for all $x \neq x_0$ in I; i.e., $f(x) - l(x) > 0$ for all $x \neq x_0$ in I; i.e., $f(x) > l(x)$ for all $x \neq x_0$ in I; i.e., the graph of f lies above the tangent line at x_0, so f is concave upward at x_0.

The case where f' is decreasing at x_0 is treated identically; just reverse all the inequality signs above.

Since f' is increasing at x_0 when $f''(x_0) > 0$, and f' is decreasing at x_0 when $f''(x_0) < 0$, we obtain the following test. (See Fig. 6-6.)

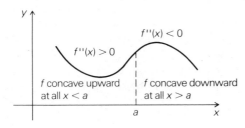

Fig. 6-6 f is concave upward (holds water) when $f'' > 0$ and is concave downward (sheds water) when $f'' < 0$.

Corollary *Let f be a differentiable function and suppose that $f''(x_0)$ exists.*

1. *If $f''(x_0) > 0$, then f is concave upward at x_0.*

2. *If $f''(x_0) < 0$, then f is concave downward at x_0.*

3. *If $f''(x_0) = 0$, then f at x_0 may be concave upward, concave downward, or neither.*

The functions x^4, $-x^4$, x^3, and $-x^3$ provide examples of all the possibilities in case 3.

Solved Exercises

6. Show:

   (a) x^4 is concave upward at zero;

   (b) $-x^4$ is concave downward at zero;

   (c) x^3 is neither concave upward nor concave downward at zero.

7. Find the points at which $f(x) = 3x^3 - 8x + 12$ is concave upward and those at which it is concave downward. Make a rough sketch of the graph.

8. (a) Relate the sign of the error made in the linear approximation to f with the second derivative of f.

    (b) Apply your conclusion to the linear approximation of $1/x$ at $x_0 = 1$.

Exercises

7. Find the intervals on which each of the following functions is concave upward or downward. Sketch their graphs.

    (a) $1/x$ (b) $1/(1 + x^2)$ (c) $x^3 + 4x^2 - 8x + 1$

8. (a) Use the second derivative to compare x^2 with $9 + 6(x - 3)$ for x near 3.

    (b) Show by algebra that $x^2 > 9 + 6(x - 3)$ for all x.

Inflection Points

We just saw that a function f is concave upward where $f''(x) > 0$ and concave downward where $f''(x) < 0$. Points which separate intervals where f has the two types of concavity are of special interest and are called *inflection points*. An example is shown in Fig. 6-7. The following is the most useful technical definition.

Definition   The point x_0 is called an *inflection point* for the function f if f' is defined and differentiable near x_0 and f'' changes sign at x_0.

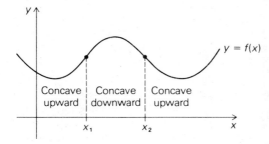

Concave   Concave   Concave
upward   downward   upward

Fig. 6-7 x_1 and x_2 are inflection points.

Comparing the preceding definition with the one on p. 74, we find that *inflection points for f are just turning points for f'*; therefore we can use all our techniques for finding turning points in order to locate inflection points. We summarize these tests in the following theorem.

Theorem 4

1. *If x_0 is an inflection point for f, then $f''(x_0) = 0$.*

2. *If $f''(x_0) = 0$ and $f'''(x_0) \neq 0$, then x_0 is an inflection point for f.*

If you draw in the tangent line to the graph of $f(x)$ at $x = a$ in Fig. 6-6, you will find that the line overtakes the graph of f at a. The opposite can occur, too; in fact, we have the following theorem, which is useful for graphing.

Theorem 5   *Suppose that x_0 is an inflection point for f.*

1. *If f'' changes from negative to positive at x_0, in particular, if $f''(x_0) = 0$ and $f'''(x_0) > 0$, then the graph of f overtakes its tangent line at x_0.*

2. *If f'' changes from positive to negative at x_0, in particular, if $f''(x_0) = 0$ and $f'''(x_0) < 0$, then the graph of f is overtaken by its tangent line at x_0.*

We invite you to try writing out the proof of Theorem 5 (see Solved Exercise 10). The two cases are illustrated in Fig. 6-8.

Recall that in defining the derivative we stipulated that all lines *other than* the tangent line must cross the graph of the function. We can now give a fairly

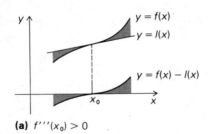

(a) $f'''(x_0) > 0$

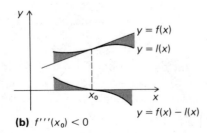

(b) $f'''(x_0) < 0$

Fig. 6-8 The tangent line crosses the graph when $f(x) - l(x)$ changes sign.

complete answer to the question of how the tangent line itself behaves with respect to the graph. Near a point x_0 where $f''(x_0) \neq 0$, the tangent line lies above or below the graph and *does not* cross it. If $f''(x_0) = 0$ but $f'''(x_0) \neq 0$, the tangent line does cross the graph. If $f''(x_0)$ and $f'''(x_0)$ are both zero, one must examine f'' on both sides of x_0 (or use $f''''(x_0)$) to analyze the situation.

Solved Exercises

9. Discuss the behavior of $f(x)$ at $x = 0$ for $f(x) = x^4, -x^4, x^5, -x^5$.

10. Prove Theorem 5. [*Hint*: See Solved Exercise 5.]

11. Find the inflection points of the function $f(x) = 24x^4 - 32x^3 + 9x^2 + 1$.

Exercises

9. Find the inflection points for x^n, n a positive integer. How does the answer depend upon n?

10. Find the inflection points for the following functions. In each case, tell how the tangent line behaves with respect to the graph.

    (a) $x^3 - x$ (b) $x^4 - x^2 + 1$

    (c) $(x - 1)^4$ (d) $1/(1 + x^2)$

11. Suppose that $f'(x_0) = f''(x_0) = f'''(x_0) = 0$, but $f''''(x_0) \neq 0$. Is x_0 a turning point or an inflection point? Give examples to show that anything can happen if $f''''(x_0) = 0$.

The General Cubic

Analytic geometry teaches us how to plot the general linear function, $f(x) = ax + b$, and the general quadratic function $f(x) = ax^2 + bx + c$ $(a \neq 0)$. The methods of this section yield the same results; the graph of $ax + b$ is a straight line, while the graph of $f(x) = ax^2 + bx + c$ is a parabola, concave upward if $a > 0$ and concave downward if $a < 0$. Moreover, the turning point of the parabola occurs when $f'(x) = 2ax + b = 0$; that is, $x = -(b/2a)$.

A more ambitious task is to determine the shape of the graph of the general cubic $f(x) = ax^3 + bx^2 + cx + d$. (We assume that $a \neq 0$; otherwise, we are dealing with a quadratic or linear function.) Of course, any specific cubic can be plotted by techniques already developed, but we wish to get an idea of what *all possible* cubics look like and how their shapes depend on a, b, c, and d.

We begin our analysis with some simplifying transformations. First of all, we can factor out a and obtain a new polynomial $f_1(x)$ as follows:

$$f(x) = a\left(x^3 + \frac{b}{a}x^2 + \frac{c}{a}x + \frac{d}{a} \right) = af_1(x)$$

The graphs of f and f_1 have the same basic shape; if $a > 0$, the y axis is just re-scaled by multiplying all y values by a; if $a < 0$, the y axis is rescaled and the graph is flipped about the x axis. It follows that we do not lose any generality by assuming that the coefficient of x^3 is 1.

Consider therefore the simpler form

$$f_1(x) = x^3 + b_1 x^2 + c_1 x + d_1$$

where $b_1 = b/a$, $c_1 = c/a$, and $d_1 = d/a$. In trying to solve cubic equations, mathematicians of the early Renaissance noticed a useful trick: if we replace x by $x - (b_1/3)$, then the quadratic term drops out; that is,

$$f_1\left(x - \frac{b_1}{3} \right) = x^3 + c_2 x + d_2$$

where c_2 and d_2 are new constants, depending on b_1, c_1, and d_1. (We leave to the reader the task of verifying this last statement and expressing c_2 and d_2 in terms of b_1, c_1, and d_1.)

The graph of $f_1(x - b_1/3) = f_2(x)$ is the same as that of $f_1(x)$ except that it is shifted by $b_1/3$ units along the x axis. This means that we lose no generality by assuming that the coefficient of x^2 is zero—that is, we only need to graph $f_2(x) = x^3 + c_2 x + d_2$. Finally, replacing $f_2(x)$ by $f_3(x) = f_2(x) - d_2$ just corresponds to shifting the graph d_2 units parallel to the y axis.

We have now reduced the graphing of the general cubic to the case of graphing $f_3(x) = x^3 + c_2 x$. For simplicity let us write $f(x)$ for $f_3(x)$ and c for c_2. To plot $f(x) = x^3 + cx$, we make the following remarks:

1. f is odd.

2. f is defined everywhere. Since $f(x) = x^3(1 - c/x^2)$, $f(x)$ is large positive [negative] when x is large positive [negative].

3. $f'(x) = 3x^2 + c$. If $c > 0$, $f'(x) > 0$ for all x, and f is increasing every-where. If $c = 0$, $f'(x) > 0$ except at $x = 0$, so f is increasing everywhere. If $c < 0$, $f'(x)$ has roots at $\pm\sqrt{-c/3}$; f is increasing on $(-\infty, -\sqrt{-c/3}]$ and $[\sqrt{-c/3}, \infty)$ and decreasing on $[-\sqrt{-c/3}, \sqrt{-c/3}]$. Thus $-\sqrt{-c/3}$ is a local maximum point and $\sqrt{-c/3}$ is a local minimum point.

4. $f''(x) = 6x$, so f is concave downward for $x < 0$ and concave upward for $x > 0$. Zero is an inflection point.

5. $f(0) = 0, f'(0) = c$

   $f(\pm\sqrt{-c}) = 0, \quad f'(\pm\sqrt{-c}) = -2c$ (if $c < 0$)

   $f\left(\pm\sqrt{\dfrac{-c}{3}}\right) = \pm\dfrac{2}{3}c\sqrt{\dfrac{-c}{3}}, \quad f'\left(\pm\sqrt{\dfrac{-c}{3}}\right) = 0$ (if $c < 0$)

Using this information, we draw the final graphs (Fig. 6-9).

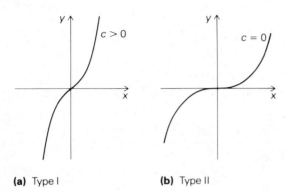

(a) Type I   **(b)** Type II

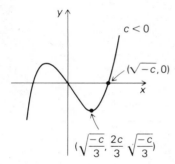

(c) Type III

Fig. 6-9 The three types of cubics.

Thus there are three types of cubics (with $a > 0$):

Type I: f is increasing at all points; $f' > 0$ everywhere.

Type II: f is increasing at all points; $f' = 0$ at one point.

Type III: f has two turning points.

Type II is the transition type between types I and III. You may imagine the graph changing as c begins with a negative value and then moves toward zero. As c gets smaller and smaller, the turning points move in toward the origin, the bumps in the graph becoming smaller and smaller until, when $c = 0$, the bumps merge at the point where $f'(x) = 0$. As c passes zero to become positive, the bumps disappear completely.

Solved Exercises

12. Convert $2x^3 + 3x^2 + x + 1$ to the form $x^3 + cx$ and determine whether the cubic is of type I, II, or III.

Exercises

12. Convert $x^3 + x^2 + 3x + 1$ to the form $x^3 + cx$ and determine whether the cubic is of type I, II, or III.

13. Convert $x^3 - 3x^2 + 3x + 1$ to the form $x^3 + cx$ and determine whether the cubic is of type I, II, or III.

14. (a) Find an explicit formula for the coefficient c_2 in $f_1(x - (b_1/3))$ in terms of b_1, c_1, and d_1, and thereby give a simple rule for determining whether the cubic $x^3 + bx^2 + cx + d$ is of type I, II, or III.

    (b) Give a rule, in terms of a, b, c, d, for determining the type of the general cubic $ax^3 + bx^2 + cx + d$.

    (c) Use the quadratic formula on the derivative of $ax^3 + bx^2 + cx + d$ to determine, in terms of a, b, c, and d, how many turning points there are. Compare with the result in (b).

The General Quartic

Having given a complete description of all cubics, we turn now to quartics:

$$f(x) = ax^4 + bx^3 + cx^2 + dx + e \quad (a \neq 0)$$

Exactly as we did with cubics, we can assume that $a = 1$, $b = 0$ (by the substitution of $x - (b/4)$ for x), and can assume that $e = 0$ (by shifting the y-axis). Thus our quartic becomes (with new coefficients!)

$$f(x) = x^4 + cx^2 + dx$$

Note that

$$f'(x) = 4x^3 + 2cx + d$$

and

$$f''(x) = 12x^2 + 2c$$

From our study of cubics, we know that $f'(x)$ has one of three shapes, according to whether $c > 0$, $c = 0$, or $c < 0$ (types I, II, and III). It turns out that there are 9 types of quartics, six of which may be divided into pairs which are mirror images of one another.

Type I $c > 0$. In this case, $f''(x)$ is everywhere positive, so f is concave upward and f' is increasing on $(-\infty, \infty)$. Thus, f' has one root, which must be a local minimum point for f. The graph looks like type I in Fig. 6-11.

Type II $c = 0$. Here the cubic $f'(x)$ is of type II and is still increasing. $f'(x) = 12x^2 \geqslant 0$, so there are no inflection points, and f is everywhere concave upward. We can subdivide type II acording to the sign of d.

Type II_1: $d > 0$. Looks like type I. (See Solved Exercise 13.)

Type II_2: $d = 0$. Looks like type I, except that the local minimum point is flatter because $f', f'', $ and f''' all vanish there.

Type II_3: $d < 0$. Looks like type I. (See Solved Exercise 13.)

The graphs are sketched in Fig. 6-11.

Type III $c < 0$. This is the most interesting case. Here, the cubic $f'(x)$ has two turning points and has 1, 2, or 3 roots depending upon its position with respect to the x axis. (Fig. 6-10.)

To determine in terms of c and d which of the five possibilities in Fig. 6-10 holds for $f'(x) = 4x^3 + 2cx + d$, we notice from the figure that it suffices to determine the sign of $f'(x)$ at the turning point of f' (which are inflection points of f). These points are obtained by solving $0 = f''(x) = 12x^2 + 2c$. They occur when $x = \pm\sqrt{-c/6}$. Now $f'(-\sqrt{-c/6}) = \frac{4}{3} c\sqrt{-c/6} + d$ is the value at the local minimum and $f'(\sqrt{-c/3}) = -\frac{4}{3}c\sqrt{-c/6} + d$ is the value at the local maximum.

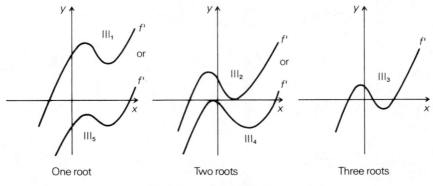

One root   Two roots   Three roots

Derivatives of type III quartics.

Fig. 6-10 Derivatives of type III quartics.

Now let's look at all the possibilities.

Type III$_1$: $0 < \frac{4}{3}c\sqrt{-c/6} + d$. Here, f' is positive at its local minimum point, so f' has one root (Fig. 6-10). From the form of f', we infer that f has the form indicated in Fig. 6-11, with one local minimum point and two inflection points.

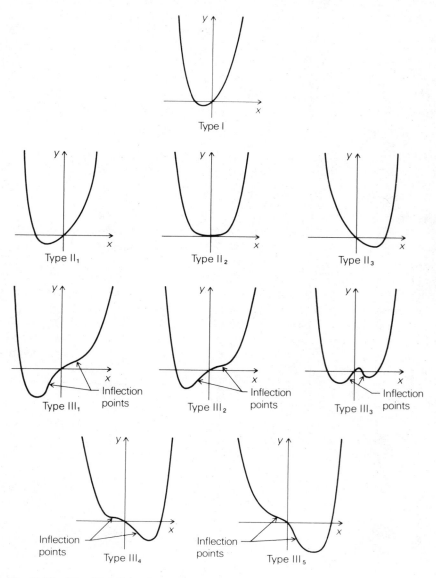

Fig. 6-11 Classification of quartics.

Type III_2: $0 = \frac{4}{3}c\sqrt{-c/6} + d$. Here, f' has two roots, and the tangent to f at one inflection point is horizontal. So f has one local minimum point and two inflection points, with horizontal tangent at one of the inflection points.

Type III_3: $\frac{4}{3}c\sqrt{-c/6} + d < 0 < \frac{4}{3}(-c)\sqrt{-c/6} + d$. Here, f' has three roots at which f'' is, from left to right, positive, negative, and positive. Thus, f has a local maximum point between two local minimum points. The three turning points are separated by two inflection points.

Type III_4: $0 = \frac{4}{3}(-c)\sqrt{-c/6} + d$. This is similar to type III_2, except that the inflection points are reversed.

Type III_5: The mirror image of type III_1.

We may draw a graph in the c, d plane (Figure 6-12) to indicate which values of c and d correspond to each type of quartic. The curve separating type I from type III is the straight line $c = 0$ (type II), while the various subtypes of type III are separated by the curve satisfying the condition (types III_2 and III_4):

$$\tfrac{4}{3}(\pm c)\sqrt{\frac{-c}{6} + d} = 0$$

i.e.,

$$-\tfrac{4}{3}c^2\left(\frac{c}{6}\right) = d^2$$

i.e.,

$$c^3 = -\tfrac{9}{2}d^2$$

This curve has a sharp point called a *cusp* at the origin.

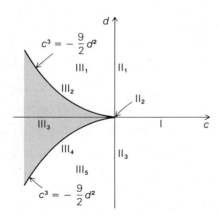

Fig. 6-12 Location of the types of quartics in the c, d plane.

Remark 1 If you rotate Fig. 6-12 90°counterclockwise, you will see that the positioning of the types corresponds to that in Fig. 6-11.

Remark 2 We have here an example of transition *curves*, as opposed to transition points. Types I, III_1, III_3, and III_5 correspond to *regions* in the c-d plane, which are separated by the transition types II_1, II_3, III_2, and III_4 lying along curves. The curves come together at the point corresponding to type II_2. In the language of catastrophe theory (see the footnote on p. 91), the function $f(x) = x^4$ of type II_2 is called the *organizing center* of the family of the quartics—in this single curve of type II_2, all the other types of curves are latent; the slightest change in c or d will destroy type II_2 and one of the other types will make its appearance.

Remark 3 We can also classify the quartics according to the number of points where the first derivative is zero and their behavior near these points, i.e., according to the number of turning points, inflection points, etc. From this point of view, the two basic types are type I-II_1-II_3-III_1-III_5 (one turning point) and type III_3 (three turning points), separated by the curve of type III_2-III_4 (one turning point and one inflection point with horizontal tangent, in which two turning points are latent). The point of type II_2, with a flat turning point, is still the organizing center.

Remark 4 Figure 6-13 shows the set of points (c, d, x) in space where x is a critical point of $f(x) = x^4 + cx^2 + dx$. (See the reference in the footnote on p. 91 for applications of this picture.)

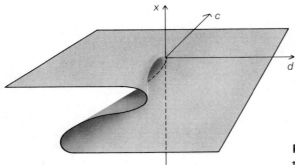

Fig. 6-13 The cusp catastrophe.

Solved Exercise

13. What geometric distinctions can you find between types I, II_1, and II_3?

Exercises

15. Classify the following quartics. Sketch their graphs.

    (a) $x^4 - 3x^2 + 2\sqrt{2}x$ (b) $x^4 - 3x^2 - 4x$

    (c) $x^4 + 4x^2 + 6x$ (d) $x^4 + 7x$

    (e) $x^4 - 6x^2 + 5x$ (f) $x^4 + x^3$

    (g) $x^4 - x^3$

16. If the equation $x^4 + bx^3 + cx^2 + dx + e = 0$ has three distinct solutions, what type of quartic must we have?

Problems for Chapter 6 ▆▆▆▆▆▆▆▆▆▆▆▆▆▆▆▆▆▆▆▆▆▆▆▆▆▆▆

1. Let f be a nonconstant polynomial such that $f(0) = f(1)$. Prove that f has a turning point somewhere in the interval $(0, 1)$.

2. What does the graph of $[ax/(bx + c)] + d$ look like, if a, b, c, and d are positive constants?

3. Prove that the graph of any cubic $f(x) = ax^3 + bx^2 + cx + d$ $(a \neq 0)$ is symmetric about its inflection point in the sense that the function

$$g(x) = f\left(x + \frac{b}{3a}\right) - f\left(-\frac{b}{3a}\right)$$

    is odd. [*Hint*: $g(x)$ is again a cubic; where is its inflection point?]

4. For the general quartic $ax^4 + bx^3 + cx^2 + dx + e$ $(a \neq 0)$, find conditions in terms of a, b, c, d, and e which determine the type of the quartic.

5. Give a classification of general *quintics* $q(x) = ax^5 + bx^4 + cx^3 + dx^2 + ex + f$ $(a \neq 0)$ analogous to the classification of quartics in the text. [*Hint*: reduce to the case $b = f = 0$ and apply the classification of quartics to the derivative $q'(x)$.] *

6. Let $f(x) = 1/(1 + x^2)$.

    (a) For which values of c is the function $f(x) + cx$ increasing on the whole real line? Sketch the graph of $f(x) + cx$ for one such c.

    (b) For which values of c is the function $f(x) - cx$ decreasing on the whole real line? Sketch the graph of $f(x) - cx$ for one such c.

    (c) How are your answers in (a) and (b) related to the inflection points of f?

7. (R. Rivlin) A rubber cube of incompressible material is pulled on all faces with a force T. The material stretches by a factor v in two directions and

---

*This is the "swallowtail catastrophe." The "butterfly catastrophe" is associated with the general sixth-degree polynomial. See T. Poston and I. Stewart, *Catastrophe Theory and Its Applications* (Fearon-Pitman (1977) for details).

contracts by a factor v^{-2} in the other. By balancing forces, one can show that

$$v^3 - \frac{T}{2\alpha}v^2 + 1 = 0 \qquad\qquad (R)$$

where α is a constant (analogous to the spring constant for a spring). Show that (R) has one (real) solution if $T < 3\sqrt[3]{2}\alpha$ and has three solutions if $T > 3\sqrt[3]{2}\alpha$.

7   The Mean Value Theorem

The mean value theorem is, like the intermediate value and extreme value theorems, an *existence* theorem. It asserts the existence of a point in an interval where a function has a particular behavior, but it does not tell you how to find the point.

Theorem 1 Mean Value Theorem. Suppose that the function f is continuous on the closed interval $[a,b]$ and differentiable on the open interval (a,b). Then there is a point x_0 in the open interval (a,b) at which $f'(x_0) = [f(b) - f(a)]/(b - a)$.

In physical terms, the mean value theorem says that the average velocity of a moving object during an interval of time is equal to the instantaneous velocity at some moment in the interval. Geometrically, the theorem says that a secant line drawn through two points on a smooth graph is parallel to the tangent line at some intermediate point on the curve. There may be more than one such point, as in Fig. 7-1. Consideration of these physical and geometric interpretations will make the theorem believable.

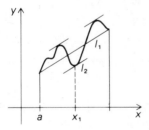

Fig. 7-1 Tangent lines parallel to the secant line.

We will prove the mean value theorem at the end of this section. For now, we will concentrate on some applications. Our first corollary tells us that if we know something about $f'(x)$ for all x in $[a,b]$, then we can conclude something about the relation between values of $f(x)$ at different points in $[a,b]$.

Corollary 1 Let f be differentiable on (a,b) [and continuous on $[a,b]$]. Suppose that, for all x in the open interval (a,b), the derivative $f'(x)$ belongs to a certain set S of real numbers. Then, for any two distinct points x_1 and x_2 in (a,b) [in $[a,b]$], the difference quotient

$$\frac{f(x_2) - f(x_1)}{x_2 - x_1}$$

belongs to S as well.

Proof The difference quotient stays the same if we exchange x_1 and x_2, so we may assume that $x_1 < x_2$. The interval $[x_1, x_2]$ is contained in (a, b) [in $[a, b]$]. Since f is differentiable on (a, b) [continuous on $[a, b]$], it is continuous on $[x_1, x_2]$ and differentiable on (x_1, x_2). By the mean value theorem, applied to f on $[x_1, x_2]$, there is a number x_0 in (x_1, x_2) such that $[f(x_2) - f(x_1)]/(x_2 - x_1) = f'(x_0)$. But (x_1, x_2) is contained in (a, b), so $x_0 \in (a, b)$. By hypothesis, $f'(x_0)$ must belong to S; hence so does $[f(x_2) - f(x_1)]/(x_2 - x_1)$.

Worked Example 1 Suppose that f is differentiable on the whole real line and that $f'(x)$ is constant. Use Corollary 1 to prove that f is linear.

Solution Let m be the constant value of f' and let S be the set whose only member is m. For any x, we may apply Corollary 1 with $x_1 = 0$ and $x_2 = x$ to conclude that $[f(x) - f(0)]/(x - 0)$ belongs to S; that is, $[f(x) - f(0)]/(x - 0) = m$. But then $f(x) = mx + f(0)$ for all x, so f is linear.

Worked Example 2 Let f be continuous on $[1, 3]$ and differentiable on $(1, 3)$. Suppose that, for all x in $(1, 3)$, $1 \leqslant f'(x) \leqslant 2$. Prove that $2 \leqslant f(3) - f(1) \leqslant 4$.

Solution Apply Corollary 1, with S equal to the interval $[1, 2]$. Then $1 \leqslant [f(3) - f(1)]/(3 - 1) \leqslant 2$, and so $2 \leqslant f(3) - f(1) \leqslant 4$.

Corollary 2 *Suppose that $f'(x) = 0$ for all x in some open interval (a, b). Then f is constant on (a, b).*

Proof Let $x_1 < x_2$ be any two points in (a, b). Corollary 1 applies with S the set consisting only of the number zero. Thus we have the equation $[f(x_2) - f(x_1)]/(x_2 - x_1) = 0$, or $f(x_2) = f(x_1)$, so f takes the same value at all points of (a, b).

Worked Example 3 Let $f(x) = d|x|/dx$

(a) Find $f'(x)$.

(b) What does Corollary 2 tell you about f? What does it not tell you?

Solution

(a) Since $|x|$ is linear on $(-\infty, 0)$ and $(0, \infty)$, its second derivative $d^2|x|/dx^2 = f'(x)$ is identically zero for all $x \neq 0$.

(b) By Corollary 2, f is constant *on any open interval on which it is differentiable*. It follows that f is constant on $(-\infty, 0)$ and $(0, \infty)$. The corollary does *not* say that f is constant on $(-\infty, \infty)$. In fact, $f(-2) = -1$, while $f(2) = +1$.

Finally, we can derive from Corollary 2 the fact that two antiderivatives of a function differ by a constant. (An antiderivative of f is a function whose derivative is f.)

Corollary 3 *Let $F(x)$ and $G(x)$ be functions such that $F'(x) = G'(x)$ for all x in an open interval (a, b). Then there is a constant C such that $F(x) = G(x) + C$ for all x in (a, b).*

Proof We apply Corollary 2 to the difference $F(x) - G(x)$. Since $(d/dx)[F(x) - G(x)] = F'(x) - G'(x) = 0$ for all x in (a, b), $F(x) - G(x)$ is equal to a constant C, and so $F(x) = G(x) + C$.

Worked Example 4 Suppose that $F'(x) = x$ for all x and that $F(3) = 2$. What is $F(x)$?

Solution Let $G(x) = \frac{1}{2}x^2$. Then $G'(x) = x = F'(x)$, so $F(x) = G(x) + C = \frac{1}{2}x^2 + C$. To evaluate C, we set $x = 3$: $2 = F(3) = \frac{1}{2}(3^2) + C = \frac{9}{2} + C$. Thus $C = 2 - \frac{9}{2} = -\frac{5}{2}$ and $F(x) = \frac{1}{2}x^2 - \frac{5}{2}$.

Solved Exercises*

1. Let $f(x) = x^3$ on the interval $[-2, 3]$. Find explicitly the value(s) of x_0 whose existence is guaranteed by the mean value theorem.

2. If, in Corollary 1, the set S is taken to be the interval $(0, \infty)$, the result is a theorem which has already been proved. What theorem is it?

3. The velocity of a train is kept between 40 and 50 kilometers per hour

---

*Solutions appear in the Appendix.

during a trip of 200 kilometers. What can you say about the duration of the trip?

4. Suppose that $F'(x) = -(1/x^2)$ for all $x \neq 0$. Is $F(x) = (1/x) + C$, where C is a constant?

_____

Exercises

1. Directly verify the validity of the mean value theorem for $f(x) = x^2 - x + 1$ on $[-1, 2]$ by finding the point(s) x_0. Sketch.

2. Suppose that f is continuous on $[0, \frac{1}{2}]$ and $0.3 \leqslant f'(x) < 1$ for $0 < x < \frac{1}{2}$. Prove that $0.15 \leqslant [f(\frac{1}{2}) - f(0)] < 0.5$.

3. Suppose that $f'(x) = x^2$ and $f(1) = 0$. What is $f(x)$?

4. Suppose that an object lies at $x = 4$ when $t = 0$ and that the velocity dx/dt is 35 with a possible error of ± 1, for all t in $[0, 2]$. What can you say about the object's position when $t = 2$?

Proof of the Mean Value Theorem

Our proof of the mean value theorem will use two results already proved which we recall here:

1. If x_0 lies in the open interval (a, b) and is a maximum or minimum point for a function f on an interval $[a, b]$ and if f is differentiable at x_0, then $f'(x_0) = 0$. This follows immediately from Theorem 3, p. 64, since if $f'(x_0)$ were not zero, f would be increasing or decreasing at x_0.

2. If f is continuous on a closed interval $[a, b]$, then f has a maximum and a minimum point in $[a, b]$ (extreme value theorem, p. 71).

The proof of the mean-value theorem proceeds in three steps.

_____

Step 1 (Rolle's Theorem). *Let f be continuous on $[a, b]$ and differentiable on (a, b), and assume that $f(a) = f(b) = 0$. Then there is a point x_0 in (a, b) at which $f'(x_0) = 0$.*

Proof If $f(x) = 0$ for all x in $[a, b]$, we can choose any x_0 in (a, b). So assume that f is not everywhere zero. By result 2 above, f has a maximum point x_1 and a minimum point x_2. Since f is zero at the ends of the interval but is not identically zero, at least one of x_1, x_2 lies in (a, b). Let x_0 be this point. By result 1, $f'(x_0) = 0$.

Rolle's theorem has a simple geometric interpretation (see Fig. 7-2).

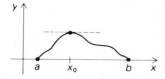

Fig. 7-2 The tangent line is horizontal at x_0.

Step 2 **(Horserace Theorem).** *Suppose that f_1 and f_2 are continuous on $[a, b]$ and differentiable on (a, b), and assume that $f_1(a) = f_2(a)$ and $f_1(b) = f_2(b)$. Then there exists a point x_0 in (a, b) such that $f_1'(x_0) = f_2'(x_0)$.*

Proof Let $f(x) = f_1(x) - f_2(x)$. Since f_1 and f_2 are differentiable on (a, b) and continuous on $[a, b]$, so is f (see Problem 6 in Chapter 5). By assumption, $f(a) = f(b) = 0$, so from step 1, $f'(x_0) = 0$ for some x_0 in (a, b). Thus $f_1'(x_0) = f_2'(x_0)$ as required.

We call this the horserace theorem because it has the following interpretation. Suppose that two horses run a race starting together and ending in a tie. Then, at some time during the race, they must have had the same velocity.

Step 3 We apply step 2 to a given function f and the linear function l that matches f at its endpoints, namely.,

$$l(x) = f(a) + (x - a) \left[\frac{f(b) - f(a)}{b - a} \right]$$

Note that $l(a) = f(a)$, $l(b) = f(b)$, and $l'(x) = [f(b) - f(a)]/(b - a)$. By step 2, $f'(x_0) = l'(x_0) = [f(b) - f(a)]/(b - a)$ for some point x_0 in (a, b).

Thus we have proved the mean value theorem:

Mean Value Theorem *If f is continuous on $[a, b]$ and is differentiable on (a, b), then there is a point x_0 in (a, b) at which*

$$f'(x_0) = \frac{f(b) - f(a)}{b - a}$$

Solved Exercises

5. Let $f(x) = x^4 - 9x^3 + 26x^2 - 24x$. Note that $f(0) = 0$ and $f(2) = 0$. Show without calculating that $4x^3 - 27x^2 + 52x - 24$ has a root somewhere strictly between 0 and 2.

6. Suppose that f is a differentiable function such that $f(0) = 0$ and $f(1) = 1$. Show that $f'(x_0) = 2x_0$ for some x_0 in $(0, 1)$.

Exercises

5. Suppose that the horses in the horserace theorem cross the finish line with equal velocities. Must they have had the same acceleration at some time during the race?

6. Let $f(x) = |x| - 1$. Then $f(-1) = f(1) = 0$, but $f'(x)$ is never equal to zero on $[-1, 1]$. Does this contradict Rolle's theorem? Explain.

Problems for Chapter 7

1. Suppose that $(d^2/dx^2)\,[f(x) - 2g(x)] = 0$. What can you say about the relationship between f and g?

2. Suppose that f and g are continuous on $[a, b]$ and that f' and g' are continuous on (a, b). Assume that

   $$f(a) = g(a) \quad \text{and} \quad f(b) = g(b)$$

   Prove that there is a number c in (a, b) such that the line tangent to the graph of f at $(c, f(c))$ is parallel to the line tangent to the graph of g at $(c, g(c))$.

3. Let $f(x) = x^7 - x^5 - x^4 + 2x + 1$. Prove that the graph of f has slope 2 somewhere between -1 and 1.

4. Find the antiderivatives of each of the following:
   (a) $f(x) = \frac{1}{2}x - 4x^2 + 21$
   (b) $f(x) = 6x^5 - 12x^3 + 15x - 11$
   (c) $f(x) = x^4 + 7x^3 + x^2 + x + 1$
   (d) $f(x) = (1/x^2) + 2x$
   (e) $f(x) = 2x(x^2 + 7)^{100}$

5. Find an antiderivative $F(x)$ for the given function $f(x)$ satisfying the given condition:

   (a) $f(x) = 2x^4$; $F(1) = 2$

   (b) $f(x) = 4 - x$; $F(2) = 1$

   (c) $f(x) = x^4 + x^3 + x^2$; $F(1) = 1$

   (d) $f(x) = 1/x^5$; $F(1) = 3$

6. If $f''(x) = 0$ on (a, b), what can you say about f?

7. (a) Let $f(x) = x^5 + 8x^4 - 5x^2 + 15$. Prove that somewhere between -1 and 0 the tangent line to the graph of f has slope -2.

   (b) Let $f(x) = 5x^4 + 9x^3 - 11x^2 + 10$. Prove that the graph of f has slope 9 somewhere between -1 and 1.

8. Let f be a polynomial. Suppose that f has a double root at a and at b. Show that $f'(x)$ has at least three roots in $[a, b]$.

9. Let f be twice differentiable on (a, b) and suppose f vanishes at three distinct points in (a, b). Prove that there is a point x_0 in (a, b) at which $f''(x_0) = 0$.

10. Use the mean value theorem to prove Theorem 5, Chapter 5 without the hypothesis that f' be continuous.

8   Inverse Functions and the Chain Rule

Formulas for the derivatives of inverse and composite functions are two of the most useful tools of differential calculus. As usual, standard calculus texts should be consulted for additional applications.

Inverse Functions

Definition Let the function f be defined on a set A. Let B be the range of values for f on A; that is, $y \in B$ means that $y = f(x)$ for some x in A. We say that f is *invertible on* A if, for every y in B, there is a *unique* x in A such that $y = f(x)$.

If f is invertible on A, then there is a function g, whose domain is B, given by this rule: $g(y)$ is that unique x in A for which $f(x) = y$. We call g the *inverse function* of f on A. The inverse function to f is denoted by f^{-1}.

Worked Example 1 Let $f(x) = x^2$. Find a suitable A such that f is invertible on A. Find the inverse function and sketch its graph.

Solution The graph of f on $(-\infty, \infty)$ is shown in Fig. 8-1. For y in the range of f—that is, $y \geqslant 0$—there are two values of x such that $f(x) = y$: namely, $-\sqrt{y}$ and $+\sqrt{y}$. To assure only one x we may restrict f to $A = [0, \infty)$. Then $B = [0, \infty)$ and for any y in B there is exactly one x in A such that $f(x) = y$: namely, $x = \sqrt{y} = g(y)$. (Choosing $A = (-\infty, 0]$ and obtaining $x = -\sqrt{y}$ is also acceptable.) We obtain the graph of the inverse by looking at the graph of f from the back of the page.

If we are given a formula for $y = f(x)$ in terms of x, we may be able to find a formula for its inverse $x = g(y)$ by solving the equation $y = f(x)$ for x in terms of y. Sometimes, for complicated functions f, one cannot solve $y = f(x)$ to get an explicit formula for x in terms of y. In that case, one must resort to theoretical results which guarantee the existence of an inverse function. These will be discussed shortly.

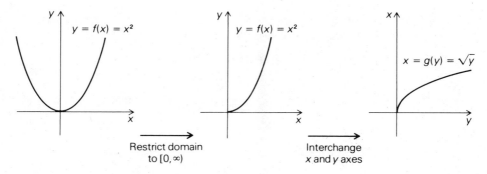

Restrict domain to [0, ∞)

Interchange x and y axes

Fig. 8-1 A restriction of $f(x) = x^2$ and its inverse.

Worked Example 2 Does $f(x) = x^3$ have an inverse on $(-\infty, \infty)$? Sketch.

Solution From Fig. 8-2 we see that the range of f is $(-\infty, \infty)$ and for each $y \in (-\infty, \infty)$ there is exactly one number x such that $f(x) = y$—namely, $x = \sqrt[3]{y} = f^{-1}(y)$ (negative if $y < 0$, positive or zero if $y \geq 0$)—so the answer is yes. We can regain x as our independent variable by observing that if $f^{-1}(y) = \sqrt[3]{y}$, then $f^{-1}(x) = \sqrt[3]{x}$. Thus we can replace y by x so that the independent variable has a more familiar name. As shown in Fig. 8-2(b) and (c), this renaming does not affect the graph of f^{-1}.

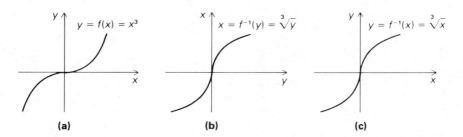

(a)   (b)   (c)

Fig. 8-2 The inverse of the cubing function is the cube root, whatever you call the variables.

Solved Exercises*

1. If $m \neq 0$, find the inverse function for $f(x) = mx + b$ on $(-\infty, \infty)$.

2. Find the inverse function for $f(x) = (ax + b)/(cx + d)$ on its domain $(c \neq 0)$.

---

*Solutions appear in the Appendix.

3. Find an inverse function g for $f(x) = x^2 + 2x + 1$ on some interval containing zero. What is $g(9)$? What is $g(x)$?

4. Sketch the graph of the inverse function for each function in Fig. 8-3.

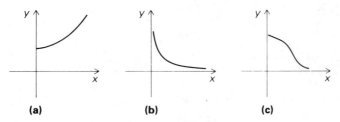

(a)   (b)   (c)

Fig. 8-3 Sketch the inverse functions.

5. Determine whether or not each function in Fig. 8-4 is invertible on its domain.

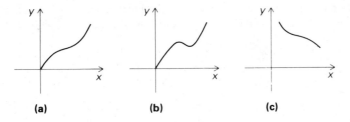

(a)   (b)   (c)

Fig. 8-4 Which functions are invertible?

___

Exercises

1. Find the inverse for each of the following functions on the given interval:
   (a) $f(x) = 2x + 5$ on $[-4, 4]$ (b) $f(x) = -\frac{1}{3}x + 2$ on $(-\infty, \infty)$
   (c) $h(t) = t - 10$ on $[0, \pi)$
   (d) $a(s) = (2s + 5)/(-s + 1)$ on $[-\frac{1}{2}, \frac{1}{2}]$
   (e) $f(x) = x^5$ on $(-\infty, \infty)$ (f) $f(x) = x^8$ on $(0, 1]$

2. Determine whether each function in Fig. 8-5 has an inverse. Sketch the inverse if there is one.

3. Sketch a graph of $f(x) = x/(1 + x^2)$ and find an interval on which f is invertible.

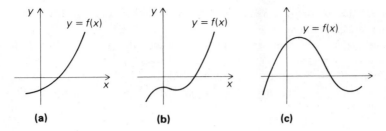

Fig. 8-5 Which functions are invertible?

4. Enter the number 2.6 on a calculator, then push the x^2 key followed by the $\sqrt{x}$ key. Is there any round-off error? Try the $\sqrt{x}$ key, then the x^2 key. Also try a sequence such as $x^2, \sqrt{x}, x^2, \sqrt{x}, \dots$. Do the errors build up? Try pushing the x^2 key five times, then the $\sqrt{x}$ key five times. Do you get back the original number? Try these experiments with different starting numbers.

5. If we think of a French-English dictionary as defining a function from the set of French words to the set of English words (does it really?), how is the inverse function defined? Discuss.

A Test for Invertibility

A function may be invertible even though we cannot find an explicit formula for the inverse function. This fact gives us a way of obtaining "new functions." There is a useful calculus test for finding intervals on which a function is invertible.

Theorem 1 *Suppose that f is continuous on* $[a,b]$ *and that f is increasing at each point of* (a,b). *(For instance, this holds if* $f'(x) > 0$ *for each x in* (a,b).) *Then f is invertible on* $[a,b]$, *and the inverse* f^{-1} *is defined on* $[f(a), f(b)]$.

If f is decreasing rather than increasing at each point of (a,b), *then f is still invertible; in this case, the domain of* f^{-1} *is* $[f(b), f(a)]$.

Proof If f is continuous on $[a,b]$ and increasing at each point of (a,b), we know by the results of Chapter 5 on increasing functions that f is increasing on $[a,b]$; that is, if $a \leqslant x_1 < x_2 \leqslant b$, then $f(x_1) < f(x_2)$. In particular, $f(a) < f(b)$. If y is any number in $(f(a), f(b))$, then by the intermediate value theorem there is an x in (a, b) such that $f(x) = y$. If $y = f(a)$ or $f(b)$, we can choose $x = a$ or $x = b$. Since f is increasing on $[a, b]$ for any y in $[f(a), f(b)]$, there can only be one x such that $y = f(x)$. In fact, if we have $f(x_1) = f(x_2) = y$, with $x_1 < x_2$, we would have $y < y$, which is impossible. Thus, by definition, f is invertible on $[a, b]$ and the domain of f^{-1} is the range $[f(a), f(b)]$ of values of f on $[a, b]$. The proof of the second assertion is similar.

Worked Example 3 Verify, using Theorem 1, that $f(x) = x^2$ has an inverse if f is defined on $[0, b]$ for a given $b > 0$.

Solution Since f is differentiable on $(-\infty, \infty)$, it is continuous on $(-\infty, \infty)$ and hence on $[0, b]$. But $f'(x) = 2x > 0$ for $0 < x < b$. Thus f is increasing. Hence Theorem 1 guarantees that f has an inverse defined on $[0, b^2]$.

In general, a function is not monotonic throughout the interval on which it is defined. Theorem 1 shows that the turning points of f divide the domain of f into subintervals on each of which f is monotonic and invertible.

Solved Exercises

6. Let $f(x) = x^5 + x$.

   (a) Show that f has an inverse on $[-2, 2]$. What is the domain of this inverse?

   (b) Show that f has an inverse on $(-\infty, \infty)$.

   (c) What is $f^{-1}(2)$?

   (d) Numerically calculate $f^{-1}(3)$ to two decimal places of accuracy.

7. Find intervals on which $f(x) = x^5 - x$ is invertible.

8. Show that, if n is odd, $f(x) = x^n$ is invertible on $(-\infty, \infty)$. What is the domain of the inverse function?

9. Discuss the invertibility of $f(x) = x^n$ for n even.

Exercises

6. Show that $f(x) = -x^3 - 2x + 1$ is invertible on $[-1, 2]$. What is the domain of the inverse?

7. (a) Show that $f(x) = x^3 - 2x + 1$ is invertible on $[2, 4]$. What is the domain of the inverse?

(b) Find the largest possible intervals on which f is invertible.

8. Find the largest possible intervals on which $f(x) = 1/(x^2 - 1)$ is invertible. Sketch the graphs of the inverse functions.

9. Show that $f(x) = \frac{1}{3}x^3 - x$ is not invertible on any open interval containing 1.

10. Let $f(x) = x^5 + x$.

   (a) Find $f^{-1}(246)$.

   (b) Find $f^{-1}(4)$, correct to at least two decimal places using a calculator.

Differentiating Inverse Functions

Even though the inverse function $f^{-1}(y)$ is defined somewhat abstractly, there is a simple formula for its derivative. To motivate the formula we proceed as follows. First of all, note that if l is the linear function $l(x) = mx + b$, then $l^{-1}(y) = (1/m)y - (b/m)$ (see Solved Exercise 1), so $l'(x) = m$, and $(l^{-1})'(y) = 1/m$, the reciprocal. We can express this by $dx/dy = 1/(dy/dx)$.*

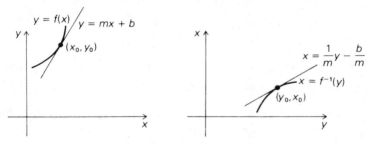

Fig. 8-6 Flipping the graphs preserves tangency.

Next, examine Fig. 8-6, where we have flipped the graph of f and its tangent line $y = mx + b$ in the drawing on the left to obtain the graph of f^{-1} together with the line $x = l^{-1}(y) = (y/m) - (b/m)$ in the drawing on the right. Since the line is tangent to the curve on the left, and flipping the drawing should preserve this tangency, the line $x = (1/m)y - (b/m)$ ought to be the tangent line to $x = f^{-1}(y)$ at (y_0, x_0), and its slope $1/m$ should be the derivative $(f^{-1})'(y_0)$. Since $m = f'(x_0)$, we conclude that

---

*As usual in calculus, we use the Leibniz notation dy/dx for $f'(x)$, when $y = f(x)$.

$$(f^{-1})'(y_0) = \frac{1}{f'(x_0)}$$

Since the expression $(f^{-1})'$ is awkward, we sometimes revert to the notation $g(y)$ for the inverse function and write

$$g'(y_0) = \frac{1}{f'(x_0)}$$

Notice that although dy/dx is not an ordinary fraction, the rule

$$\frac{dx}{dy} = \frac{1}{dy/dx}$$

is valid.

Theorem 2 *Suppose that $f'(x) > 0$ or $f'(x) < 0$ for all x in an open interval I containing x_0, so that by Theorem 1 there is an inverse function g to f, defined on an open interval containing $f(x_0) = y_0$, with $g(y_0) = x_0$. Then g is differentiable at y_0 and*

$$g'(y_0) = \frac{1}{f'(x_0)} = \frac{1}{f'(g(y_0))}$$

Proof We will consider the case $f'(x) > 0$. The case $f'(x) < 0$ is entirely similar (in fact, one may deduce the case $f' < 0$ from the case $f' > 0$ by considering $-f$).

From Theorem 1, f is increasing on I and the domain of g is an open interval J. The graphs of typical f and g are drawn in Fig. 8-7, which you should consult while following the proof.

Let $x = l(y)$ be a line through $(y_0, g(y_0)) = (y_0, x_0)$ (Fig. 8-7) with slope $m > 1/f'(x_0) > 0$. We will prove that l overtakes g at y_0. We may write $l(y) = my + b$, so $l(y_0) = x_0$. Define $\tilde{l}(x) = (x/m) - (b/m)$ which is the inverse function of $l(y)$. The line $y = \tilde{l}(x)$ passes through $(x_0, f(x_0))$ and has slope $1/m < f'(x_0)$.

By definition of the derivative, $\tilde{l}$ is overtaken by f at x_0. This means that there is an interval I containing x_0 such that

$$f(x) > \tilde{l}(x) \quad \text{if } x \text{ is in } I_1, x > x_0$$

and

$$f(x) < \tilde{l}(x) \quad \text{if } x \text{ is in } I_1, x < x_0$$

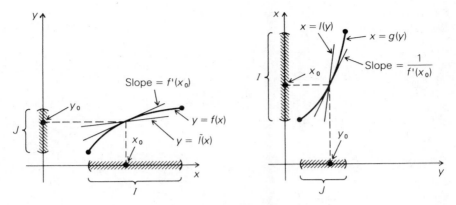

Fig. 8-7 Illustrating the proof of the inverse function theorem.

Let J_1 be the corresponding y interval, an open interval containing y_0 (see Theorem 1). Since f is increasing, $x > x_0$ corresponds to $y > y_0$ if $y = f(x)$, and likewise $x < x_0$ corresponds to $y < y_0$.

Now let y be in J_1, $y > y_0$, and let $y = f(x)$ so $x > x_0$ and x is in I_1. Hence

$$f(x) > \bar{l}(x) = \frac{x}{m} - \frac{b}{m}$$

Since $y = f(x)$ and $x = g(y)$, this becomes

$$y > \frac{g(y)}{m} - \frac{b}{m}$$

i.e.,

$$l(y) = my + b > g(y)$$

since $m > 0$. Similarly, one shows that if $y < y_0$, and y is in I,

$$l(y) < g(y)$$

Hence l overtakes g at y_0, which proves what we promised.

A similar argument shows that a line with slope $m < 1/f'(x_0)$ is overtaken by g at y_0. Thus, by definition of the derivative, g is differentiable at y_0, and its derivative there is $1/f'(x_0)$ as required.

Worked Example 4 Use the inverse function rule to compute the derivative of $\sqrt{x}$. Evaluate the derivative at $x = 2$.

Solution Let us write $g(y) = \sqrt{y}$. This is the inverse function to $f(x) = x^2$. Since $f'(x) = 2x$,

$$g'(y) = \frac{1}{f'(g(y))} = \frac{1}{2g(y)} = \frac{1}{2\sqrt{y}}$$

so $(d/dy)\sqrt{y} = 1/(2\sqrt{y})$. We may substitute any letter for y in this result, including x, so we get the formula

$$\frac{d}{dx}\sqrt{x} = \frac{1}{2\sqrt{x}}$$

When $x = 2$, the derivative is $1/(2\sqrt{2})$.

Generalizing this example, we can show that if $f(x) = x^r$, r a rational number, then

$$f'(x) = rx^{r-1}$$

(First consider the case where $1/r$ is an integer.)

Solved Exercises

10. Verify the inverse function rule for $y = (ax + b)/(cx + d)$ by finding dy/dx and dx/dy directly. (See Solved Exercise 2.)

11. If $f(x) = x^3 + 2x + 1$, show that f has an inverse on $[0,2]$. Find the derivative of the inverse function at $y = 4$.

12. Give an example of a differentiable increasing function $f(x)$ on $(-\infty, \infty)$ which has an inverse $g(y)$, but such that g is not differentiable at $y = 0$.

13. Check the self-consistency of the inverse function theorem in this sense: if it is applied to the inverse of the inverse function, we recover something we know to be true.

Exercises

11. Let $y = x^3 + 2$. Find dx/dy when $y = 3$.

12. If $f(x) = x^5 + x$, find the derivative of the inverse function when $y = 34$.

13. Find the derivative of $\sqrt[3]{x}$ on $I = (0, \infty)$.

14. For each function f below, find the derivative of the inverse function g at the points indicated:

(a) $f(x) = 3x + 5$; find $g'(2), g'(\frac{3}{4})$.

(b) $f(x) = x^5 + x^3 + 2x$; find $g'(0), g'(4)$.

(c) $f(x) = \frac{1}{12}x^3 - x$ on $[-1, 1]$; find $g'(0), g'(\frac{11}{12})$.

15. Carry out the details of the proof of Theorem 2 in the case $m < 1/f'(x_0)$. [*Hint*: First consider the case $m > 0$.]

Composition of Functions

The idea behind the definition of composition of functions is that one variable depends on another through an intermediate one.

Definition Let f and g be functions with domains D_f and D_g. Let D be the set consisting of those x in D_g for which $g(x)$ belongs to D_f. For x in D, we can evaluate $f(g(x))$, and we call the result $h(x)$. The resulting function $h(x) = f(g(x))$, with domain D, is called the *composition* of f and g. It is often denoted by $f \circ g$.

Worked Example 5 If $f(u) = u^3 + 2$ and $g(x) = \sqrt{x^2 + 1}$, what is $h = f \circ g$?

Solution We calculate $h(x) = f(g(x))$ by writing $u = g(x)$ and substituting in $f(u)$. We get $u = \sqrt{x^2 + 1}$ and

$$h(x) = f(u) = u^3 + 2 = (\sqrt{x^2 + 1})^3 + 2 = (x^2 + 1)^{3/2} + 2$$

Since each of f and g is defined for all numbers, the domain D of h is $(-\infty, \infty)$.

Solved Exercises

14. Let $g(x) = x + 1$ and $f(u) = u^2$. Find $f \circ g$ and $g \circ f$.

15. Let $h(x) = x^{24} + 3x^{12} + 1$. Write $h(x)$ as a composite function $f(g(x))$.

16. Let $f(x) = x - 1$ and $g(x) = \sqrt{x}$.

    (a) What are the domains D_f and D_g?

    (b) Find $f \circ g$ and $g \circ f$. What are their domains?

    (c) Find $(f \circ g)(2)$ and $(g \circ f)(2)$.

    (d) Sketch graphs of $f, g, f \circ g$, and $g \circ f$.

17. Let i be the identity function $i(x) = x$. Show that $i \circ f = f$ and $f \circ i = f$ for any function f.

Exercises

16. Find $f \circ g$ and $g \circ f$ in each of the following cases.

    (a) $g(x) = x^3$; $f(x) = \sqrt{x - 2}$

    (b) $g(x) = x^r$; $f(x) = x^s$ (r, s rational)

    (c) $g(x) = 1/(1 - x)$; $f(x) = \frac{1}{2} - \sqrt{3x}$

    (d) $g(x) = (3x - 2)/(4x + 1)$; $f(x) = (2x - 7)/(9x + 3)$

17. Write the following as compositions of simpler functions:

    (a) $h(x) = 4x^2/(x^2 - 1)$

    (b) $h(r) = (r^2 + 6r + 9)^{3/2} + (1/\sqrt{r^2 + 6r + 9})$

    (c) $h(u) = \sqrt{(1 - u)/(1 + u)}$

18. Show that the inverse f^{-1} of any function f satisfies $f \circ f^{-1} = i$ and $f^{-1} \circ f = i$, where i is the identity function of Solved Exercise 17.

The Chain Rule

Theorem 3   Chain Rule. *Suppose that g is differentiable at x_0 and that f is differentiable at $g(x_0)$. Then $f \circ g$ is differentiable at x_0 and its derivative there is $f'(g(x_0)) \cdot g'(x_0)$.*

As with the algebraic rules of differentiation, to prove the chain rule, we reduce the problem to one about rapidly vanishing functions (See Chapter 3). Write $y_0 = g(x_0)$ and

$$f(y) = f(y_0) + f'(y_0)(y - y_0) + r(y)$$

where $r(y)$ is rapidly vanishing at y_0. Set $y = g(x)$ in this formula to get

$$f(g(x)) = f(g(x_0)) + f'(g(x_0))(g(x) - g(x_0)) + r(g(x))$$

Suppose we knew that $r(g(x))$ vanished rapidly at x_0. Then we could differentiate the right-hand side with respect to x at $x = x_0$ to obtain what we want (using the sum rule, the constant multiple rule, and the fact that the derivative of a constant is zero). In other words, the proof of Theorem 3 is reduced to the following result.

Lemma *Let $r(y)$ be rapidly vanishing at y_0 and suppose that $g(x)$ is differentiable at x_0, where $y_0 = g(x_0)$. Then $r(g(x))$ is rapidly vanishing at x_0.*

Proof We will use the characterization of rapidly vanishing functions given by Theorem 1 of Chapter 3. First of all, we have $r(g(x_0)) = r(y_0) = 0$, since r vanishes at y_0. Now, given any number $\epsilon > 0$, we must find an open interval I about x_0 such that, for all x in I with $x \neq x_0$, we have $|r(g(x))| < \epsilon |x - x_0|$.

Pick a number M such that $M > |g'(x_0)|$. Since r vanishes rapidly at y_0 and ϵ/M is a positive number, there is an open interval J about y_0 such that, for all y in J with $y \neq y_0$, we have $|r(y)| < (\epsilon/M) |y - y_0|$.

Next, let I be an open interval about x_0 such that, for $x \neq x_0$ in I, we have $|g(x) - g(x_0)| < M|x - x_0|$. Such an interval exists because $-M < g'(x_0) < M$. (See Fig. 8-8.) By making I sufficiently small (see Solved Exercise 19), we can be sure that $g(x)$ is in J for all x in I.

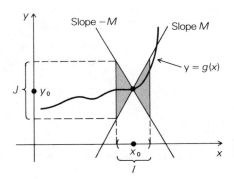

Fig. 8-8 The graph $y = g(x)$ lies in the "bow-tie" region for x in I.

Finally, let $x \in I$, $x \neq x_0$. To show that $|r(g(x))| < \epsilon |x - x_0|$, we consider separately the cases $g(x) \neq y_0$ and $g(x) = y_0$. In the first case, we have

$$|r(g(x))| < \frac{\epsilon}{M}|g(x) - y_0| = \frac{\epsilon}{M}|g(x) - g(x_0)| < \frac{\epsilon}{M} \cdot M|x - x_0| = \epsilon |x - x_0|$$

In the second case, $r(g(x)) = 0$ and $\epsilon |x - x_0| > 0$, so we are done.

Worked Example 6 Verify the chain rule for $f(u) = u^2$ and $g(x) = x^3 + 1$.

Solution Let $h(x) = f(g(x)) = [g(x)]^2 = (x^3 + 1)^2 = x^6 + 2x^3 + 1$. Thus $h'(x) = 6x^5 + 6x^2$. On the other hand, since $f'(u) = 2u$ and $g'(x) = 3x^2$,

$$f'(g(x)) \cdot g'(x) = (2 \cdot (x^3 + 1)) \, 3x^2 = 6x^5 + 6x^2$$

Hence the chain rule is verified in this case.

Solved Exercises

18. Show that the rule for differentiating $1/g(x)$ follows from that for $1/x$ and the chain rule.

19. How should we choose I in the proof of the lemma so that $g(x)$ is in J for all x in I?

Exercises

19. Suppose f is differentiable on (a, b) with $f'(x) > 0$ for all x, and g is differentiable on (c, d) with $g'(x) < 0$ for x in (c, d). Let (c, d) be the image set of f, so $g \circ f$ is defined. Show that $g \circ f$ has an inverse and calculate the derivative of the inverse.

20. Let f, g, h all be differentiable. State a theorem concerning the differentiability of $k(x) = h(g(f(x)))$. Think of a *specific* example where you would apply this result.

21. Prove that $\dfrac{d}{dx} [f(x)]^r = r [f(x)]^{r-1} f'(x)$ (rational power of a function rule).

Problems for Chapter 8

1. Prove that, if f is continuous and increasing (or decreasing) on an open interval I, then the inverse function f^{-1} is continuous as well.

2. Let f be differentiable on an open interval I. Assume f' is continuous and $f'(x_0) \neq 0$. Prove that, on an interval about x_0, f has a differentiable inverse g.

3. Suppose that f is differentiable on an open interval I and that f' is continuous. Assume $f'(x) \neq 0$ for all x in I. Prove f has an inverse which is differentiable.

4. Let $f^{(n)}(x) = f \circ \cdots \circ f$ (n times). Express $f^{(n)'}(0)$ in terms of $f'(0)$ if $f(0) = 0$.

5. Show that the inverse function rule can be derived from the chain rule if you assume that the inverse function is differentiable (use the relation $f^{-1}(f(x)) = x$).

6. Let $f(x) = x^3 - 3x + 7$.

   (a) Find an interval containing zero on which f is invertible. Denote the inverse by g.

   (b) What is the domain of g?

   (c) Calculate $g'(7)$.

7. Let $f(x) = (ax + b)/(cx + d)$, and let $g(x) = (rx + s)/(tx + u)$.

   (a) Show that $f \circ g$ and $g \circ f$ are both of the form $(kx + l)/(mx + n)$ for some $k, l, m,$ and n.

   (b) Under what conditions on a, b, c, d, r, s, t, u does $f \circ g = g \circ f$?

8. If f is a given differentiable function and $g(x) = f(\sqrt{x})$, what is $g'(x)$?

9. If f is differentiable and has an inverse, and $g(x) = f^{-1}(\sqrt{x})$, what is $g'(x)$?

10. Find a formula for the second derivative of $f \circ g$ in terms of the first and second derivatives of f and g.

11. Find a formula for the second derivative of $g(y)$ if $g(y)$ is the inverse function of $f(x)$.

12. Derive a formula for differentiating the composition $f_1 \cdot f_2 \cdot \ldots \cdot f_n$ of n functions.

9 The Trigonometric Functions

The theory of the trigonometric functions depends upon the notion of arc length on a circle, in terms of which radian measure is defined. It is possible to develop this theory from scratch, using the integral (just as for the logarithm), but intuition is sacrificed in this approach. At the expense of some rigor, we shall take for granted the properties of arc lengths of circular arcs and the attendant properties of π and the trigonometric functions.

The Derivative of sin θ and cos θ

We recall that if an arc length θ is measured along the unit circle in the x, y plane counterclockwise from $(1,0)$ to a point P, then the coordinates of P are $(\cos\theta, \sin\theta)$. (See Fig. 9-1.) The measure of the angle swept out is θ *radians*. The graphs of $\sin\theta$ and $\cos\theta$ are shown in Fig. 9-2.

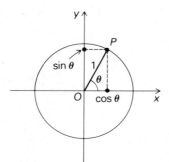

Fig. 9-1 The definition of $\cos\theta$ and $\sin\theta$.

The key to differentiating the trigonometric functions is the following lemma.

Lemma *For* $-\pi/2 < \theta < \pi/2, \theta \neq 0$, *we have*

$$1 - \frac{\theta^2}{2} < \cos\theta < \frac{\sin\theta}{\theta} < 1$$

Proof Refer to Fig. 9-3. Observe that for $0 < \theta < \pi/2$,

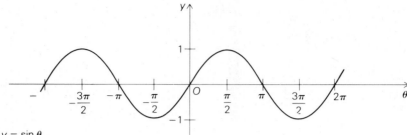

(a) $y = \sin \theta$

(b) $y = \cos \theta$

Fig. 9-2 Graphs of sine and cosine.

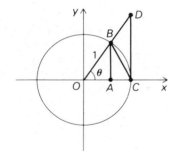

Fig. 9-3 Geometry used to determine $\sin'(0)$.

$$\text{area } \triangle OCB = \tfrac{1}{2}|OC| \cdot |AB| = \tfrac{1}{2}\sin \theta$$

and

$$\text{area } \triangle OCB < \text{area sector } OCB = \tfrac{1}{2}\theta$$

and

$$\text{area sector } OCB < \text{area } \triangle OCD = \tfrac{1}{2}|OC| \cdot |CD| = \tfrac{1}{2}\tan \theta.$$

Thus,

$$\frac{\sin\theta}{\theta} < 1 \quad \text{and} \quad \theta < \frac{\sin\theta}{\cos\theta}$$

and so

$$\cos\theta < \frac{\sin\theta}{\theta} < 1$$

This inequality also holds for $\theta < 0$, since $\cos\theta$ and $(\sin\theta)/\theta$ are both unchanged if θ is replaced by $-\theta$.

Using the identity $\cos\theta = 1 - 2\sin^2(\theta/2)$, we get $\cos\theta > 1 - (\theta^2/2)$ and hence the lemma.

___

Theorem 1   *We have*

$$\sin'(0) = 1$$

and

$$\cos'(0) = 0$$

Proof   For $m > 1$ we have

$$\frac{\sin\theta}{\theta} < 1 < m$$

and so

$$\sin\theta < m\theta \quad \text{if } \theta > 0$$

and

$$\sin\theta > m\theta \quad \text{if } \theta < 0$$

i.e., for $m > 1$, the line $y = m\theta$ (in the θ, y plane) overtakes $y = \sin\theta$ at $\theta = 0$.

Since $1 - (\theta^2/2)$ is continuous and equals 1 at $\theta = 0$, if $m < 1$, there is an interval I about 0 such that $1 - (\theta^2/2) > m$ if $\theta \in I$. Thus

$$\theta \in I \quad \text{and} \quad \theta > 0 \quad \text{implies} \quad \sin\theta > m\theta$$

and

$$\theta \in I \quad \text{and} \quad \theta < 0 \quad \text{implies} \quad \sin\theta < m\theta$$

i.e., for $m < 1$ the line $y = m\theta$ is overtaken by $y = \sin\theta$. Therefore, by definition of the derivative (Chapter 1), $\sin'(0) = 1$.

Rewriting the inequality $1 - (\theta^2/2) < \cos\theta < 1$ as $-\theta^2/2 < \cos\theta - 1 < 0$ and arguing in a similar manner, we get $\cos'(0) = 0$.

The formulas for the derivatives of $\sin\theta$ and $\cos\theta$ at a general point follow from the addition formulas and the chain rule:

$$\sin\theta = \sin[\theta_0 + (\theta - \theta_0)]$$
$$= \sin\theta_0\cos(\theta - \theta_0) + \cos\theta_0\sin(\theta - \theta_0)$$

By the chain rule and Theorem 1, the right-hand side is differentiable in θ at $\theta = \theta_0$ with derivative

$$\sin\theta_0\cos'(0) + \cos\theta_0\sin'(0) = \cos\theta_0$$

Thus

$$\sin'\theta_0 = \cos\theta_0$$

In a similar way one shows that $\cos'\theta_0 = -\sin\theta_0$. We have proved the following result.

Theorem 2 $\sin\theta$ *and* $\cos\theta$ *are differentiable functions of* θ *with*

$$\sin'\theta = \cos\theta, \quad \cos'\theta = -\sin\theta$$

The rules of calculus now enable one to differentiate expressions involving $\sin\theta$ and $\cos\theta$.

Worked Example 1 Differentiate $(\sin 3x)/(1 + \cos^2 x)$.

Solution By the chain rule,

$$\frac{d}{dx}\sin 3x = 3\cos 3x$$

So, by the quotient rule,

$$\frac{d}{dx}\frac{\sin 3x}{1 + \cos^2 x} = \frac{(1 + \cos^2 x)\cdot 3\cos 3x - \sin 3x \cdot 2\cos x(-\sin x)}{(1 + \cos^2 x)^2}$$
$$= \frac{3\cos 3x(1 + \cos^2 x) + 2\cos x\sin x\cdot\sin 3x}{(1 + \cos^2 x)^2}$$

Now that we know how to differentiate the sine and cosine functions, we can differentiate the remaining trigonometric functions by using the rules of calculus. For example, the quotient rule gives:

$$\frac{d}{d\theta} \tan \theta = \frac{\cos \theta \, (d/d\theta) \sin \theta - \sin \theta \, (d/d\theta) \cos \theta}{\cos^2 \theta}$$

$$= \frac{\cos \theta \cdot \cos \theta + \sin \theta \cdot \sin \theta}{\cos^2 \theta}$$

$$= \frac{1}{\cos^2 \theta} = \sec^2 \theta$$

In a similar way, we see that

$$\frac{d}{d\theta} \cot \theta = -\csc^2 \theta$$

Writing $\csc \theta = 1/\sin \theta$ we get

$$\csc' \theta = (-\sin' \theta)/(\sin^2 \theta) = (-\cos \theta)/(\sin^2 \theta) = -\cot \theta \csc \theta$$

and similarly

$$\sec' \theta = \tan \theta \sec \theta$$

Worked Example 2 Differentiate $\csc x \tan 2x$.

Solution Using the product rule and the chain rule,

$$\frac{d}{dx} \csc x \tan 2x = \left(\frac{d}{dx} \csc x\right)(\tan 2x) + \csc x \left(\frac{d}{dx} \tan 2x\right)$$

$$= -\cot x \cdot \csc x \cdot \tan 2x + \csc x \cdot 2 \cdot \sec^2 2x$$

$$= 2 \csc x \sec^2 2x - \cot x \cdot \csc x \cdot \tan 2x$$

Solved Exercises*

1. Differentiate:
   (a) $\sin x \cos x$ \qquad (b) $(\tan 3x)/(1 + \sin^2 x)$ \qquad (c) $1 - \csc^2 5x$

2. Differentiate $f(\theta) = \sin(\sqrt{3\theta^2 + 1})$.

3. Discuss maxima, minima, concavity, and points of inflection for $f(x) = \sin^2 x$. Sketch its graph.

---

*Solutions appear in the Appendix.

Exercises

1. Show that $(d/d\theta)\cos\theta = -\sin\theta$ can be derived from $(d/d\theta)\sin\theta = \cos\theta$ by using $\cos\theta = \sin(\pi/2 - \theta)$ and the chain rule.

2. Differentiate:
   (a) $\sin^2 x$

   (b) $\tan(\theta + 1/\theta)$
   (c) $(4t^3 + 1)\sin\sqrt{t}$

   (d) $\csc t \cdot \sec^2 3t$

3. Discuss maxima, minima, concavity, and points of inflection for $y = \cos 2x - 1$.

The Inverse Trigonometric Functions

In Chapter 8 we discussed the general concept of the inverse of a function and developed a formula for differentiating the inverse. Recall that this formula is

$$\frac{d}{dy}f^{-1}(y) = \frac{1}{(d/dx)f(x)}$$

where $y = f(x)$.

To apply these ideas to the sine function, we begin by using Theorem 1 (of Chapter 8) to locate an interval on which $\sin x$ has an inverse. Since $\sin' x = \cos x > 0$ on $(-\pi/2, \pi/2)$, $\sin x$ is increasing on this interval, so $\sin x$ has an inverse on the interval $[-\pi/2, \pi/2]$. The inverse is denoted $\sin^{-1} y$.* We obtain the graph of $\sin^{-1} y$ by interchanging the x and y coordinates. (See Fig. 9-4.)

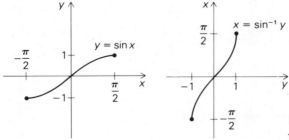

Fig. 9-4 The sine function and its inverse.

*Although the notation $\sin^2 y$ is commonly used to mean $(\sin y)^2$, $\sin^{-1} y$ does not mean $(\sin y)^{-1} = 1/\sin y$. Sometimes the notation arcsin y is used for the inverse sine function to avoid confusion.

The values of $\sin^{-1} y$ may be obtained from a table for $\sin x$. (Many pocket calculators can evaluate the inverse trigonometric functions as well as the trigonometric functions.)

Worked Example 3 Calculate $\sin^{-1} 1$, $\sin^{-1} 0$, $\sin^{-1} (-1)$, $\sin^{-1} (-\frac{1}{2})$, and $\sin^{-1} (0.342)$.

Solution Since $\sin \pi/2 = 1$, $\sin^{-1} 1 = \pi/2$. Similarly, $\sin^{-1} 0 = 0$, $\sin^{-1} (-1) = -\pi/2$. Also, $\sin(-\pi/6) = -\frac{1}{2}$, so $\sin^{-1} (-\frac{1}{2}) = -\pi/6$. Using a calculator, we find $\sin^{-1} (0.342) = 20°$.

We could have used any other interval on which $\sin x$ has an inverse, such as $[\pi/2, 3\pi/2]$, to define an inverse sine function; had we done so, the function obtained would have been different. The choice $[-\pi/2, \pi/2]$ is standard and is the most convenient.

Let us now calculate the derivative of $\sin^{-1} y$. By the formula for the derivative of an inverse,

$$\frac{d}{dy} \sin^{-1} y = \frac{1}{(d/dx) \sin x} = \frac{1}{\cos x}$$

where $y = \sin x$. However, $\cos^2 x + \sin^2 x = 1$, so $\cos x = \sqrt{1 - y^2}$. (The negative root does not occur since $\cos x$ is positive on $(-\pi/2, \pi/2)$.)

Thus

$$\frac{d}{dy} \sin^{-1} y = \frac{1}{\sqrt{1 - y^2}} = (1 - y^2)^{-1/2}, \quad -1 < y < 1$$

Notice that the derivative of $\sin^{-1} y$ is not defined at $y = \pm 1$ but is "infinite" there. This is consistent with the appearance of the graph in Fig. 9-4.

Worked Example 4 Differentiate $h(y) = \sin^{-1} (3y^2)$.

Solution From the chain rule, with $u = 3y^2$,

$$h'(y) = (1 - u^2)^{-1/2} \frac{du}{dy} = 6y(1 - 9y^4)^{-1/2}$$

Worked Example 5 Differentiate $f(x) = x \sin^{-1} (2x)$.

Solution Here we are using x for the variable name. Of course we can use any letter we please. By the product and chain rules,

$$f'(x) = \left(\frac{dx}{dx}\right) \sin^{-1} (2x) + x \frac{d}{dx} (\sin^{-1} 2x)$$

$$= \sin^{-1} 2x + 2x(1 - 4x^2)^{-1/2}$$

It is interesting to observe that, while $\sin^{-1} y$ is defined in terms of trigonometric functions, its derivative is an algebraic function, even though the derivatives of the trigonometric functions themselves are still trigonometric.

Solved Exercises

4. (a) Calculate $\sin^{-1} (\frac{1}{2})$, $\sin^{-1} (-\sqrt{3}/2)$, and $\sin^{-1} (2)$.
   (b) Simplify $\tan (\sin^{-1} x)$.

5. Calculate $(d/dx)(\sin^{-1} 2x)^{3/2}$.

6. Differentiate $\sin^{-1} (\sqrt{1 - x^2})$, $0 < x < 1$. Discuss.

Exercises

4. What are $\sin^{-1} (0.3)$, $\sin^{-1} (2/\sqrt{3})$, $\sin^{-1} (\frac{3}{2})$, $\sin^{-1} (-\pi)$, and $\sin^{-1} (1/\sqrt{3})$?

5. Differentiate the indicated functions:
   (a) $(x^2 - 1) \sin^{-1} (x^2)$ (b) $(\sin^{-1} x)^2$
   (c) $\sin^{-1} [t/(t + 1)]$ (domain = ?)

6. What are the maxima, minima, and inflection points of $f(x) = \sin^{-1} x$?

7. Show that $\cos x$ on $(0, \pi)$ has an inverse $\cos^{-1} x$ and

$$\frac{d}{dx} \cos^{-1} x = -\frac{1}{\sqrt{1 - x^2}}$$

8. Show that $\tan^{-1} x$ is defined for all x, takes values between $-\pi/2$ and $\pi/2$, and

$$\frac{d}{dx} \tan^{-1} x = \frac{1}{1 + x^2}$$

Problems for Chapter 9 ▬▬▬▬▬

1. Differentiate each of the following functions:
   (a) $f(x) = \sqrt{x} + \cos 3x$ (b) $f(x) = \sqrt{\cos x}$
   (c) $f(x) = (\sin^{-1} 3x)/(x^2 + 2)$ (d) $f(x) = (x^2 \cos^{-1} x + \tan x)^{3/2}$
   (e) $f(\theta) = \cot^{-1} (\sin \theta + \sqrt{\cos^2 3\theta + \theta^2})$
   (f) $f(r) = (r^2 + \sqrt{1 - r^2})/(r \sin r)$

2. Differentiate $\sec [\sin^{-1} (y - 2)]$ by (a) simplifying first and (b) using the chain rule right away.

3. (a) What is the domain of $\cos^{-1}(x^2 - 3)$? Differentiate.

   (b) Sketch the graph of $\cos^{-1}(x^2 - 3)$.

4. Show that $f(x) = \sec x$ satisfies the equation $f'' + f - 2f^3 = 0$.

5. Is the following correct: $(d/dx)\cos^{-1} x = (-1)(\cos^{-2} x)[(d/dx)\cos x]$?

6. Find a function $f(x)$ which is differentiable and increasing for all x, yet $f(x) < \pi/2$ for all x.

7. Find the inflection points of $f(x) = \cos^2 3x$.

8. Where is $f(x) = x \sin x + 2 \cos x$ concave upward? Concave downward?

9. Prove that $y = \tan^{-1} x$ has an inflection point at $x = 0$.

10. A child is whirling a stone on a string 0.5 meter long in a vertical circle at 5 revolutions per second. The sun is shining directly overhead. What is the velocity of the stone's shadow when the stone is at the 10 o'clock position?

11. Prove that $f(x) = x - 1 - \cos x$ is increasing on $[0, \infty)$. What inequality can you deduce?

10 The Exponential and Logarithm Functions

Some texts define e^x to be the inverse of the function $\ln x = \int_1^x 1/t\,dt$. This approach enables one to give a quick definition of e^x and to overcome a number of technical difficulties, but it is an unnatural way to define exponentiation. Here we give a complete account of how to define $\exp_b(x) = b^x$ as a continuation of rational exponentiation. We prove that $\exp_b$ is differentiable and show how to introduce the number e.

Powers of a Number

If n is a positive integer and b is a real number, the power b^n is defined as the product of b with itself n times:

$$b^n = b \cdot b \cdot \ldots \cdot b \;\; (n \text{ times})$$

If b is unequal to 0, so is b^n, and we define

$$b^{-n} = \frac{1}{b^n} = \frac{1}{b} \; \ldots \; \frac{1}{b} \;\; (n \text{ times}).$$

We also set

$$b^0 = 1$$

If b is positive, we define $b^{1/2} = \sqrt{b}$, $b^{1/3} = \sqrt[3]{b}$, etc., since we know how to take roots of numbers. Recall that $\sqrt[n]{b}$ is the unique positive number such that $(\sqrt[n]{b})^n = b$; i.e., $\sqrt[n]{y}$ is the inverse function of x^n. Formally, for n a positive integer, we define

$$b^{1/n} = \sqrt[n]{b} \;\; (\text{the positive } n\text{th root of } b)$$

and we define

$$b^{-1/n} = \frac{1}{b^{1/n}}$$

Worked Example 1 Express $9^{-1/2}$ and $625^{-1/4}$ as fractions.

Solution $9^{-1/2} = 1/9^{1/2} = 1/\sqrt{9} = \frac{1}{3}$ and $625^{-1/4} = 1/\sqrt[4]{625} = \frac{1}{5}$.

Worked Example 2 Show that, if we assume the rule $b^{x+y} = b^x b^y$, we are *forced* to define $b^0 = 1$ and $b^{-x} = 1/b^x$.

Solution If we set $x = 1$ and $y = 0$, we get $b^{1+0} = b^1 \cdot b^0$, i.e., $b = b \cdot b^0$ so $b^0 = 1$. Next, if we set $y = -x$, we get $b^{x-x} = b^x b^{-x}$, i.e., $1 = b^0 = b^x b^{-x}$, so $b^{-x} = 1/b^x$. (Notice that this is an argument for *defining* b^0, $b^{-1/n}$, and b^{-n} the way we did. It does not prove it. Once powers are defined, and only then, can we claim that rules like $b^{x+y} = b^x b^y$ are true.)

Finally, if r is a rational number, we define b^r by expressing r as a quotient m/n of positive integers and defining

$$b^r = (b^m)^{1/n}$$

We leave it to the reader (Exercise 8) to verify that the result is independent of the way in which r is expressed as a quotient of integers. Note that $b^{m/n}$ is always positive, even if m or n is negative.

Thus the laws of exponents,

$$b^n b^m = b^{n+m} \quad \text{and} \quad b^n/b^m = b^{n-m} \tag{i}$$
$$(b^n)^m = b^{nm} \tag{ii}$$
$$(bc)^n = b^n c^n \tag{iii}$$

which are easily seen for integer powers from the definition of power, may now be extended to rational powers.

Worked Example 3 Prove (i) for rational exponents, namely,

$$b^{m_1/n_1} \, b^{m_2/n_2} = b^{(m_1/n_1)+(m_2/n_2)} \tag{i'}$$

Solution From (iii) we get

$$(b^{m_1/n_1} \, b^{m_2/n_2})^{n_1 n_2} = (b^{m_1/n_1})^{n_1 n_2}(b^{m_2/n_2})^{n_1 n_2}$$

By (ii) this equals

$$((b^{m_1/n_1})^{n_1})^{n_2}((b^{m_2/n_2})^{n_2})^{n_1}$$

By definition of $b^{m/n}$, we have $(b^{m/n})^n = b^m$, so the preceding expression is

$$(b^{m_1})^{n_2}(b^{m_2})^{n_1} = b^{m_1 n_2} \, b^{m_2 n_1}$$

again by (ii), which equals

$$b^{m_1 n_2 + m_2 n_1}$$

by (i).

Hence

$$(b^{m_1/n_1} \, b^{m_2/n_2})^{n_1 n_2} = b^{m_1 n_2 + m_2 n_1}$$

so

$$b^{m_1/n_1} \, b^{m_2/n_2} = (b^{(m_1 n_2 + m_2 n_1)})^{1/n_1 n_2}$$

By the definition $b^{m/n} = (b^m)^{1/n}$, this equals

$$b^{(m_1 n_2 + m_2 n_1)/n_1 n_2} = b^{(m_1/n_1) + (m_2/n_2)}$$

as required.

Similarly, we can prove (ii) and (iii) for rational exponents:

$$(b^{m_1/n_1})^{m_2/n_2} = b^{m_1 m_2 / n_1 n_2} \tag{ii'}$$
$$(bc)^{m/n} = b^{m/n} c^{m/n} \tag{iii'}$$

Worked Example 4   Simplify $(x^{2/3}(x^{-3/2}))^{8/3}$.

Solution $(x^{2/3} x^{-3/2})^{8/3} = (x^{(2/3)-(3/2)})^{8/3} = (x^{-5/6})^{8/3} = x^{-20/9} = 1/\sqrt[9]{x^{20}}$.

Worked Example 5   If $b > 1$ and p and q are rational numbers with $p < q$, prove that $b^p < b^q$.

Solution   By the laws of exponents, $b^q/b^p = b^{q-p}$. Let $z = q - p > 0$. We shall show that $b^z > 1$, so $b^q/b^p > 1$ and thus $b^q > b^p$.

Suppose that $z = m/n$. Then $b^z = (b^m)^{1/n}$. However, $b^m = b \cdot b \cdot \ldots \cdot b$ (m times) > 1 since $b > 1$, and $(b^m)^{1/n} > 1$ since $b^m > 1$. (The nth root $c^{1/n}$ of a number $c > 1$ is also greater than 1, since, if $c^{1/n} \leqslant 1$, then $(c^{1/n})^n = c \leqslant 1$ also.) Thus $b^z > 1$ if $z > 0$, and the solution is complete.

As a consequence, we can say that if $b \geqslant 1$ and $p \leqslant q$, then $b^p \leqslant b^q$.

Solved Exercises*

1. Find $8^{-3/2}$ and $8^{1/2}$.

---

*Solutions appear in the Appendix.

2. Find $9^{3/2}$.

3. Simplify $(x^{2/3})^{5/2}/x^{1/4}$.

4. Verify **(ii)** if either p or q is zero.

Exercises

1. Simplify by writing with rational exponents:

   (a) $\left[\dfrac{\sqrt[4]{ab^3}}{\sqrt{b}}\right]^6$

   (b) $\sqrt[3]{\dfrac{\sqrt{a^3b^9}}{\sqrt[4]{a^6b^6}}}$

2. Factor (i.e., write in the form $(x^a \pm y^b)(x^c \pm y^d)$, a, b, c, d rational numbers):

   (a) $x - \sqrt{xy} - 2y$

   (b) $x - y$

   (c) $\sqrt[3]{xy^2} + \sqrt[3]{yx^2} + x + y$

   (d) $x - 2\sqrt{x} - 8$

   (e) $x + 2\sqrt{3x} + 3$

3. Solve for x:

   (a) $10^x = 0.001$

   (b) $5^x = 1$

   (c) $2^x = 0$

   (d) $x - 2\sqrt{x} - 3 = 0$ (factor)

4. Do the following define the same function on (a) $(-1, 1)$, (b) $(0, 3)$?

$$f_1(x) = x^{1/2}$$
$$f_2(x) = \sqrt[4]{x^2}$$
$$f_3(x) = (\sqrt[4]{x})^2 \quad \text{(which, if any, are the same?)}$$

5. Based on the laws of exponents which we want to hold true, what would be your choice for the value of 0^0? Discuss.

6. Using rational exponents and the laws of exponents, verify the following root formulas.

   (a)
   $$\sqrt[a]{\sqrt[b]{x}} = \sqrt[ab]{x}$$

   (b)
   $$\sqrt[ac]{x^{ab}} = \sqrt[c]{x^b}$$

7. Find all real numbers x which satisfy the following inequalities.

   (a) $x^{1/3} > x^{1/2}$

   (b) $x^{1/2} > x^{1/3}$

   (c) $x^{1/p} > x^{1/q}$, p, q positive odd integers, $p > q$

   (d) $x^{1/q} > x^{1/p}$, p, q positive odd integers, $p > q$

8. Suppose that $b > 0$ and that $p = m/n = m'/n'$. Show, using the definition of rational powers, that $b^{m/n} = b^{m'/n'}$; i.e., b^p is unambiguously defined.

The Function $f(x) = b^x$

Having defined $f(x) = b^x$ if x is rational, we wish to extend the definition to allow x to range through all real numbers. If we take, for example, $b = 2$ and compute some values, we get:

x	-2	$-\frac{3}{2}$	-1	$-\frac{1}{2}$	0	$\frac{1}{2}$	1	$\frac{3}{2}$	2
2^x	0.25	0.354...	0.5	0.707...	1	1.414...	2	2.828...	4

These values may be plotted to get an impression of the graph (Fig. 10-1). It seems natural to conjecture that the graph can be filled in with a smooth curve, i.e., that b^x makes sense for all x.

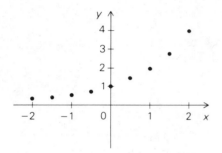

Fig. 10-1   The plot of some points $(x, 2^x)$ for rational x.

To calculate a number like $2^{\sqrt{3}}$, we should be able to take a decimal approximation to $\sqrt{3} \approx 1.732050808$ ..., say, 1.7320, calculate the rational power $2^{1.7320} = 2^{17320/10000}$, and hope to get an approximation to $2^{\sqrt{3}}$. Experimentally, this leads to reasonable data. On a calculator, one finds the following:

x	2^x
1	2
1.7	3.24900958
1.73	3.31727818
1.732	3.32188010
1.73205	3.32199523
1.7320508	3.32199707
1.732050808	3.32199708

The values of 2^x as x gets closer to $\sqrt{3}$ seem to be converging to some definite number. By doing more and more calculations, we could approximate this number to as high a degree of accuracy as we wished. We thus have a method for generating the decimal expansion of a number which could be called $2^{\sqrt{3}}$. To define $2^{\sqrt{3}}$ and other irrational powers, we shall use the transition idea.

Let b be positive and let x be irrational. Let A be the set of all real numbers α which are less than or equal to b^p, where p is some rational number and $p < x$. Similarly, let B be the set of numbers $\beta \geq b^q$ where q is some rational number and $q > x$ (Fig. 10-2).

Fig. 10-2 Powers of 2 with rational exponents less than $\sqrt{3}$ go into set A (along with all numbers less than such powers) and all powers of 2 with rational exponents larger than $\sqrt{3}$ go into set B (along with all numbers larger than such powers).

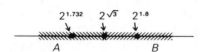

$2^{1.732}$ $2^{\sqrt{3}}$ $2^{1.8}$

A B

Theorem 1   *There is exactly one transition point from A to B if $b > 1$ (and from B to A if $0 < b < 1$). This transition point is called b^x (if $b = 1$, we define $1^x = 1$ for all x).*
    The function b^x so obtained is a continuous function of x.

The proof is given in the next two sections. (We shall assume it for now.) There we shall also show that the laws of exponents for rational numbers remain true for arbitrary real exponents. A specific case follows.

Worked Example 6   Simplify $(\sqrt{(3^\pi)})(3^{-\pi/4})$.

Solution $\sqrt{3^\pi}\, 3^{-\pi/4} = (3^\pi)^{1/2} 3^{-\pi/4} = 3^{(\pi/2)-(\pi/4)} = 3^{\pi/4}$.

Sometimes the notation $\exp_b x$ is used for b^x, exp standing for "exponential." One reason for this is typographical: an expression like

$$\exp_b\left(\frac{x^2}{2} + 3x\right)$$

is easier on the eyes and on the printer than $b^{(x^2/2)+3x}$. Another reason is mathematical: when we write $\exp_b x$, we indicate that we are thinking of b^x *as a function of x*.

We saw in Worked Example 5 that, for $b > 1$ and p and q rational with $p < q$ we had $b^p < b^q$. We can prove the same thing for real exponents: if $x < y$, we can choose rational numbers, p and q, such that $x < p < q < y$. By the definition of b^x and b^y as transition points, we must have $b^x \leqslant b^p$ and $b^q \leqslant b^y$, so $b^x \leqslant b^p < b^q \leqslant b^y$, and thus $b^x < b^y$.

In functional notation, if $b > 1$, we have $\exp_b x < \exp_b y$ whenever $x < y$; in the language of Chapter 5, $\exp_b$ is an increasing function. Similarly, if $0 < b < 1$, $\exp_b$ is a decreasing function.

It follows from Theorem 1 of Chapter 8 that for $b > 1$, b^x has a unique inverse function with domain $(0, \infty)$ and range $(-\infty, \infty)$. This function is denoted $\log_b$. Thus $x = \log_b y$ is the number such that $b^x = y$.

Worked Example 7 Find $\log_3 9, \log_{10}(10^a)$, and $\log_9 3$.

Solution Let $x = \log_3 9$. Then $3^x = 9$. Since $3^2 = 9$, x must be 2. Similarly, $\log_{10} 10^a$ is a and $\log_9 3 = \frac{1}{2}$ since $9^{1/2} = 3$.

The graph of $\log_b x$ for $b > 1$ is sketched in Fig. 10-3 and is obtained by flipping over the graph of $\exp_b x$ along the diagonal $y = x$. As usual with inverse functions, the label y in $\log_b y$ is only temporary to stress the fact that $\log_b y$ is the inverse of $y = \exp_b x$. From now on we shall usually use the variable name x and write $\log_b x$.

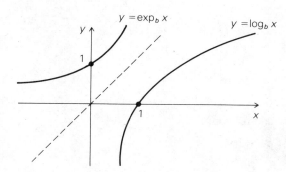

Fig. 10-3 The graphs of $y = \exp_b x$ and $y = \log_b x$ compared.

Notice that for $b > 1$, $\log_b x$ is increasing. If $b < 1$, $\exp_b x$ is decreasing and so is $\log_b x$. However, while $\exp_b x$ is always positive, $\log_b x$ can be either positive or negative.

From the laws of exponents we can read off corresponding laws for $\log_b x$:

$$\log_b (xy) = \log_b x + \log_b y \text{ and } \log_b \left(\frac{x}{y}\right) = \log_b x - \log_b y \qquad \text{(i)}$$

$$\log_b (x^y) = y \log_b x \qquad \text{(ii)}$$

$$\log_b x = \log_b (c) \log_c (x) \qquad \text{(iii)}$$

For instance, to prove **(i)**, we remember that $\log_b x$ is the number such that $\exp_b (\log_b x) = x$. So we must check that

$$\exp_b (\log_b x + \log_b y) = \exp_b (\log_b xy).$$

But the left-hand side is $\exp_b (\log_b x) \exp_b (\log_b y) = xy$ as is the right-hand side. The other laws are proved in the same way.

Solved Exercises

5. How is the graph of $\exp_{1/b} x$ related to that of $\exp_b x$?

6. Simplify: $(2^{\sqrt{3}} + 2^{-\sqrt{3}})(2^{\sqrt{3}} - 2^{-\sqrt{3}})$.

7. Match the graphs and functions in Fig. 10-4.

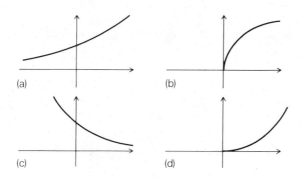

(a)   (b)   (c)   (d)

Fig. 10-4 Match the graphs and functions:

    (a) $y = x^{\sqrt{3}}$ (b) $y = x^{1/\sqrt{3}}$
    (c) $y = (\sqrt{3})^x$ (d) $y = (1/\sqrt{3})^x$

8. Find $\log_2 4$, $\log_3 81$ and $\log_{10} 0.01$.

9. (a) Simplify $\log_b (b^{2x}/2b)$

   (b) Solve for x: $\log_2 x = \log_2 5 + 3 \log_2 3$.

Exercises

9. Simplify: $\dfrac{(\sqrt{3})^{\pi} - (\sqrt{2})^{\sqrt{5}}}{\sqrt[4]{3^{\pi}} + 2^{\sqrt{5}/4}}$

10. Give the domains and ranges of the following functions and graph them:

    (a) $y = 2^{(x^2)}$ (b) $y = 2^{\sqrt{x}}$ (c) $y = 2^{1/x}$

11. Graph $y = 3^{x+2}$ by "shifting" the graph of $y = 3^x$ two units to the left. Graph $y = 9(3^x)$ by "stretching" the graph of $y = 3^x$ by a factor of 9 in the y-axis direction. Compare the two results. In general, how does shifting the graph $y = 3^x$ by k units to the left compare with stretching the graph by a factor of 3^k in the y-axis direction?

12. Consider $f(x) = (-3)^x$. For which fractions x is $f(x)$ defined? Not defined? How might this affect your ability to define $f(\pi)$?

13. Graph the following functions on one set of axes.

    (a) $f(x) = 2^x$ (b) $g(x) = x^2 + 1$ (c) $h(x) = x + 1$

    Can you make an estimate of $f'(1)$?

14. Solve for x:

    (a) $\log_x 5 = 0$ (b) $\log_2(x^2) = 4$ (c) $2 \log_3 x + \log_3 4 = 2$

15. Use the definition of $\log_b x$ to prove:

    (a) $\log_b (x^y) = y \log_b x$ (b) $\log_b x = \log_b (c) \log_c (x)$

Convex Functions*

We shall use the following notion of convexity to prove Theorem 1.

Definition Let $f(x)$ be a function defined for every rational [real] x. We call f *convex* provided that for every pair of rational [real] numbers x_1 and x_2 with $x_1 \leqslant x_2$, and rational [real] λ with $0 \leqslant \lambda \leqslant 1$ we have

---

*See "To e Via Convexity" by H. Samelson, *Am. Math. Monthly*, November 1974, p. 1012. Some valuable remarks were also given us by Peter Renz.

$$f(\lambda x_1 + (1 - \lambda)x_2) \leqslant \lambda f(x_1) + (1 - \lambda)f(x_2)$$

If $\leqslant$ can be replaced by $<$ throughout, we say that f is *strictly convex*.

Notice that $\lambda x_1 + (1 - \lambda)x_2$ lies between x_1 and x_2; for example, if $\lambda = \frac{1}{2}$, $\lambda x_1 + (1 - \lambda)x_2 = \frac{1}{2}(x_1 + x_2)$ is the midpoint. Thus convexity says that at any point z between x_1 and x_2, $(z, f(z))$ lies beneath the chord joining $(x_1, f(x_1))$ to $(x_2, f(x_2))$. (See Fig. 10-5.) To see this, notice that the equation of the chord is

$$y - f(x_1) = \frac{f(x_2) - f(x_1)}{x_2 - x_1}(x - x_1)$$

Setting $x = z = \lambda x_1 + (1 - \lambda)x_2$, we get

$$
\begin{aligned}
y &= f(x_1) + \left(\frac{f(x_2) - f(x_1)}{x_2 - x_1}\right)(\lambda x_1 + (1 - \lambda)x_2 - x_1) \\
&= f(x_1) + (f(x_2) - f(x_1))(1 - \lambda) \\
&= \lambda f(x_1) + (1 - \lambda)f(x_2)
\end{aligned}
$$

So the condition in the definition says exactly that $f(z) \leqslant \lambda f(x_1) + (1 - \lambda)f(x_2)$, the y value of the point on the chord above z.

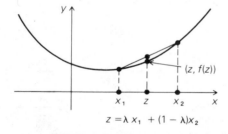

Fig. 10-5 For a convex function, the chord lies above the graph.

Theorem 2 *If $b > 1, f(x) = b^x$ defined for x rational, is (strictly) convex.*

Proof First of all, we prove that for $x_1 < x_2$,

$$f\left(\frac{x_1 + x_2}{2}\right) < \tfrac{1}{2}(f(x_1) + f(x_2))$$

i.e.,

$$b^{(x_1 + x_2)/2} < \tfrac{1}{2}(b^{x_1} + b^{x_2})$$

Indeed, this is the same as

$$b^{(x_1+x_2)/2} - b^{x_1} < b^{x_2} - b^{(x_1+x_2)/2}$$

i.e.,

$$b^{x_1}(b^{(x_2-x_1)/2} - 1) < (b^{(x_2-x_1)/2} - 1)b^{(x_1+x_2)/2}$$

Since $x_1 < (x_1 + x_2)/2$, this is indeed true, as $b^{x_1} < b^{(x_1+x_2)/2}$ (see Worked Example 5).

Having taken care of $\lambda = \frac{1}{2}$, we next assume $0 < \lambda < \frac{1}{2}$. Proceeding as above,

$$b^{\lambda x_1 + (1-\lambda)x_2} < \lambda b^{x_1} + (1 - \lambda)b^{x_2}$$

is the same as

$$\lambda(b^{\lambda x_1 + (1-\lambda)x_2} - b^{x_1}) < (1 - \lambda)(b^{x_2} - b^{\lambda x_1 + (1-\lambda)x_2})$$

i.e.,

$$\lambda b^{x_1}(b^{(1-\lambda)(x_2-x_1)} - 1) < (1 - \lambda)(b^{(1-\lambda)(x_2-x_1)} - 1)b^{\lambda x_2 + (1-\lambda)x_1}$$

But if $0 < \lambda < \frac{1}{2}$, then $\lambda < (1 - \lambda)$, and since $x_1 < \lambda x_2 + (1 - \lambda)x_1$, $b^{x_1} < b^{\lambda x_2 + (1-\lambda)x}$. Hence the inequality is true. If we replace λ by $(1 - \lambda)$ everywhere in this argument we get the desired inequality for $\frac{1}{2} < \lambda < 1$.

One can prove that b^x is convex for $b < 1$ in exactly the same manner.

Note The inequality obtained in Solved Exercise 11 is important and will be used in what follows.

Solved Exercises

10. (a) Prove that $y = x^2$ is strictly convex.

    (b) Find a convex function that is not strictly convex.

11. Suppose f is convex and $x_1 < x_2 < x_3$. Show that

$$f(x_1) \geqslant f(x_2) + \{[f(x_3) - f(x_2)]/(x_3 - x_2)\}(x_1 - x_2)$$

Sketch. What if f is strictly convex? Make up a similar inequality of the form $f(x_3) \geqslant$ something.

Exercises

16. If $f(x)$ is twice differentiable on $(-\infty, \infty)$ with $f''(x)$ continuous and $f''(x) > 0$, prove that f is convex. [*Hint:* Consider $g(x) = \lambda f(x) + (1 - \lambda)f(x_1) - f(\lambda x + (1 - \lambda)x_1)$ and show that g is increasing, and $g(x_1) = 0$.]

17. Show that $f(x) = |x|$ is convex.

Proof of Theorem 1

Let us suppose that $b > 1$ and that x is a given irrational number (the case $b < 1$ is dealt with similarly). Let A be the set of α such that $\alpha \leqslant b^p$, where p is rational and $p < x$, and let B be the set of $\beta \geqslant b^q$, where q is rational and $q > x$.

Lemma 1 *A and B are convex and hence intervals;* $A = (-\infty, \alpha_0)$ *or* $(-\infty, \alpha_0]$ *and* $B = (\beta_0, \infty)$ *or* $[\beta_0, \infty)$ *for some* $\alpha_0 \leqslant \beta_0$.

Proof Suppose y_1 and y_2 belong to A, and $y_1 < y < y_2$. Thus $y_2 \leqslant b^p$ for some rational $p < x$. Hence $y \leqslant b^p$ for the same p, so y belongs to A. This shows A is convex and, if y_2 belongs to A and $y < y_2$, then y belongs to A. Hence $A = (-\infty, \alpha_0)$ or $(-\infty, \alpha_0]$ for some number α_0.

Similarly, $B = (\beta_0, \infty)$ or $[\beta_0, \infty)$ for some number β_0.

Either $\alpha_0 \leqslant \beta_0$ or $\alpha_0 > \beta_0$. If $\alpha_0 > \beta_0$, then β_0 belongs to A so $\beta_0 \leqslant b^p$ for some $p < x$. This implies that b^p belongs to B, so $b^p \geqslant b^q$ for some $q > x$. But if $p < x < q$, $b^p < b^q$. Thus $\alpha_0 > \beta_0$ is impossible, so it must be that $\alpha_0 \leqslant \beta_0$, as required.

The next step uses convexity.

Lemma 2 *The numbers* α_0 *and* β_0 *given in Lemma 1 are equal.*

Proof Suppose $\alpha_0 < \beta_0$, the only possibility other than $\alpha_0 = \beta_0$ (see Lemma 1). Pick p and q rational with $p < x$ and $q > x$. Then b^p belongs to A, so $b^p \leqslant \alpha_0$ and similarly $b^q \geqslant \beta_0$. Picking a smaller p and larger q will insure that $b^p < \alpha_0$ and $b^q > \beta_0$.

If we choose λ such that λ is rational and $0 < (b^q - \beta_0)/(b^q - b^p) < \lambda < (b^q - \alpha_0)/(b^q - b^p) < 1$, then we will have

$$\alpha_0 < \lambda b^p + (1 - \lambda)b^q < \beta_0$$

(Why is it possible to choose such a λ?)

Suppose that $\lambda p + (1 - \lambda)q > x$. By Solved Exercise 11 with $x_1 = p$, $x_2 = \lambda p + (1 - \lambda)q$, and $x_3 = q$, we get

$$b^p \geqslant b^{\lambda p + (1-\lambda)q} + \frac{(b^q - b^{\lambda p + (1-\lambda)q}}{\lambda(q - p)}(1 - \lambda)(p - q)$$

i.e.,

$$\lambda b^p + (1 - \lambda)b^q \geqslant b^{\lambda p + (1-\lambda)q} \geqslant \beta_0$$

which is impossible, since $\lambda b^p + (1 - \lambda)b^q < \beta_0$. Similarly, if $\lambda p + (1 - \lambda)q < x$, the inequality (also from Solved Exercise 11),

$$b^q \geqslant b^{\lambda p + (1-\lambda)q} + \frac{b^p - b^{\lambda p + (1-\lambda)q}}{\lambda(p - q)}(1 - \lambda)(q - p)$$

leads to a contradiction. Since λ, p, q are rational and x is irrational, we cannot have $\lambda p + (1 - \lambda)q = x$. Hence $\alpha_0 < \beta_0$ is impossible.

Lemmas 1 and 2 can be summarized as follows: The sets A and B are as shown in Fig. 10-6 and the endpoints may or may not belong to A or B. This means that $\alpha_0 = \beta_0$ is the transition point from A to B. Thus, b^x is defined. From the construction, note that if $p < x < q$ and p, q are rational, then $b^p < b^x < b^q$.

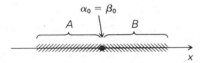

Fig. 10-6 The configuration of A, B, α_0, and β_0.

Lemma 3 *The function b^x is increasing $(b > 1)$.*

Proof If $x_1 < x_2$ and x_1, x_2 are irrational, then pick a rational r with $x_1 < r < x_2$. Then $b^{x_1} < b^r < b^{x_2}$ (see the comment just before the statement of the lemma). If x_1, x_2 are rational, see Worked Example 5.

Lemma 4 *The laws of exponents hold for b^x.*

Proof We prove $b^x b^y = b^{x+y}$. The rest are similar.

Assume $b^x b^y > b^{x+y}$, and let $\epsilon = b^x b^y - b^{x+y}$. Pick a rational number $r > x + y$ such that $b^r - b^{x+y} < \epsilon$ (why is this possible?). Write $r = p_1 + p_2$ where $p_1 > x$ and $p_2 > y$. Since the laws of exponents are true for rationals, we get

$$b^r = b^{p_1} b^{p_2} > b^x b^y$$

Hence

$$b^x b^y < b^r < b^{x+y} + \epsilon = b^{x+y} + (b^x b^y - b^{x+y}) = b^x b^y,$$

which is a contradiction. Similarly, $b^x b^y < b^{x+y}$ is impossible, so we must have equality.

Lemma 5 b^x is a (strictly) convex function (defined for every real x).

Proof Since we know that b^x is increasing and that the laws of exponents hold, our proof given in Theorem 2 is valid for arbitrary x_1, x_2, and λ, rational or not.

It only remains to prove that b^x is continuous. The following might surprise you.

Theorem 3 Any convex function $f(x)$ (defined for all real x) is continuous.

Proof Fix a number x_0 and let $c > f(x_0)$. Refer to Fig. 10-7 and the definitions on pp. 54 and 31. Pick $x_1 < x_0 < x_2$.

By convexity,

$$f(\lambda x_2 + (1 - \lambda)x_0) \leqslant \lambda f(x_2) + (1 - \lambda)f(x_0)$$

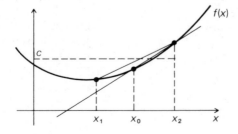

Fig. 10-7 The geometry needed for Theorem 3.

Choose ϵ_2 such that $0 < \epsilon_2 < 1$ and such that

$$\epsilon_2 f(x_2) + (1 - \epsilon_2) f(x_0) < c$$

i.e.,

$$\epsilon_2(f(x_2) - f(x_0)) < c - f(x_0)$$

(if $f(x_2) - f(x_0) < 0$, any ϵ_2 will do; if $f(x_2) - f(x_0) > 0$, we need $\epsilon_2 < (c - f(x_0))/[f(x_2) - f(x_0)]$). Then if $0 \leqslant \lambda \leqslant \epsilon_2$, $f(\lambda x_2 + (1 - \lambda)x_0) \leqslant \lambda f(x_2) + (1 - \lambda)f(x_0) < c$. If $x_0 \leqslant x \leqslant x_0 + \epsilon_2(x_2 - x_0)$, we can write $x = \lambda x_2 + (1 - \lambda)x_0$, where $\lambda = (x - x_0)/(x_2 - x_0) \leqslant \epsilon_2$. Thus

$$f(x) = f(\lambda x_2 + (1 - \lambda)x_0) \leqslant \lambda f(x_2) + (1 - \lambda)f(x_0) < c$$

whenever $x_0 \leqslant x \leqslant x_0 + \epsilon_2(x_2 - x_0)$. Similarly, by considering the line through $(x_1, f(x_1))$ and $(x_0, f(x_0))$ we can find ϵ_1 such that if $x_0 - \epsilon_1(x_0 - x_1) < x < x_0$, then $f(x) < c$. If $I = (x_0 - \epsilon_1(x_0 - x_1), x_0 + \epsilon_2(x_2 - x_0))$, then for any x in I, $f(x) < c$.

If $d < f(x_0)$, we can show that if $x_1 < x < x_0$ but x is sufficiently close to x, then $f(x) > d$ by using the inequality

$$f(x) \geqslant f(x_0) + \frac{f(x_2) - f(x_0)}{x_2 - x_0}(x - x_0)$$

and an argument like the one just given. The case $x > x_0$ is similar. Thus there is an open interval J about x_0 such that $f(x) > d$ if x is in J.

Thus, by the definition of continuous function, f is continuous at x_0.

Solved Exercises

12. Suppose that $f(x)$ is convex, $a < b$, and $f(a) < f(x)$ for every x in $(a, b]$. Prove that f is increasing on $[a, b]$.

13. A certain function $f(x)$ defined on $(-\infty, \infty)$ satisfies $f(xy) = (f(y))^x$ for all real numbers x and y. Show that $f(x) = b^x$ for some $b > 0$.

Exercises

18. Give the details of the part of the proof of Theorem 3 dealing with the case $d < f(x_0)$.

19. Prove that if f is strictly convex and $f(0) < 0$, then the equation $f(x) = 0$ has at least one real root.

20. Suppose that f satisfies $f(x + y) = f(x)f(y)$ for all real numbers x and y. Suppose that $f(x) \neq 0$ for some x. Prove the following.

(a) $f(0) = 1$

(b) $f(x) \neq 0$ for all x

(c) $f(x) > 0$ for all x

(d) $f(-x) = \dfrac{1}{f(x)}$

Differentiation of the Exponential Function

Now we turn our attention to the differentiability of b^x. Again, convexity will be an important tool.

Theorem 4   *If $b > 0$, then $f(x) = b^x$ is differentiable and*

$$f'(x) = f'(0)f(x), \quad \text{i.e.,} \quad \frac{d}{dx} b^x = f'(0) \cdot b^x$$

Proof   From the equation $f(x) = f(x - x_0)f(x_0)$ and the chain rule, all we need to do is show that $f(x)$ is differentiable at $x = 0$.

Refer to Fig. 10-8. Let $x_2 > 0$ and consider the line through $(0, 1)$ and $(x_2, f(x_2))$, i.e.,

$$y = 1 + \frac{f(x_2) - 1}{x_2} \cdot x = l_2(x)$$

This line overtakes $f(x)$ at $x = 0$. Indeed, if $0 < x < x_2$, then $f(x) < l_2(x)$ since f is (strictly) convex. If $x < 0 < x_2$, then $f(x) > l_2(x)$ by Solved Exercise 11, with $x_1, 0, x_2$ replacing x_1, x_2, and x_3.

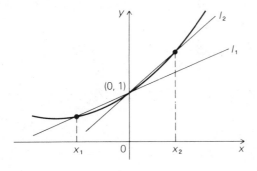

Fig. 10-8 The geometry needed for Theorem 4.

In exactly the same way, we see that for $x_1 < 0$, the line l_1 passing through $(x_1, f(x_1))$ and $(0, 1)$ is overtaken by f at $x = 0$.

Going back to the definition of derivative in terms of transitions (Theorem 4, p. 27), we let

A = the set of slopes of lines l_1 which are overtaken by f at $x = 0$

and

B = the set of slopes of lines l_2 which overtake f at $x = 0$

Let $A = (-\infty, \alpha)$ or $(-\infty, \alpha]$ and $B = (\beta, \infty)$ or $[\beta, \infty)$. We know that $\alpha \leqslant \beta$. We want to prove that $\alpha = \beta$.

Our remarks above on convexity imply that the slope of the line $y = 1 + \{[f(x_2) - 1]/x_2\}x$ belongs to B, i.e.,

$$\beta \leqslant \frac{f(x_2) - 1}{x_2}$$

Similarly,

$$\frac{f(x_1) - 1}{x_1} \leqslant \alpha$$

In particular, set $x_1 = -x_2 = -t$, where $t > 0$. Then

$$\beta - \alpha \leqslant \frac{f(x_2) - 1}{x_2} - \frac{f(x_1) - 1}{x_1} = \frac{b^t - 1}{t} - \frac{b^{-t} - 1}{-t}$$

$$= \frac{b^t + b^{-t} - 2}{t} = \frac{b^{-t}}{t}(b^{2t} - 2b^t + 1)$$

$$= \frac{b^{-t}}{t}(b^t - 1)^2$$

Now we may use the convexity inequality which tells us that $f(t) < l_2(t)$ if $0 < t < x_2$, i.e.,

$$b^t < 1 + \frac{b^{x_2} - 1}{x_2} t, \quad \text{i.e.,} \quad \frac{b^t - 1}{t} < \frac{b^{x_2} - 1}{x_2}$$

This gives

$$\beta - \alpha < b^{-t}(b^t - 1) \cdot \left(\frac{b^{x_2} - 1}{x_2}\right)$$

Suppose that $\beta - \alpha$ is positive. Then, letting $c = (\beta - \alpha)x_2/(b^{x_2} - 1)$, we have

$$b^{-t}(b^t - 1) > c$$

But $g(t) = b^{-t}(b^t - 1)$ is continuous, and $g(0) = 0$. Thus if t is near enough to zero we would have $b^{-t}(b^t - 1) < c$, a contradiction. Thus $\beta = \alpha$ and so $f'(0)$ exists.

We still need to find $f'(0)$. It would be nice to be able to adjust b so that $f'(0) = 1$, for then we would have simply $f'(x) = f(x)$. To be able to keep track of b, we revert to the $\exp_b (x)$ notation, so Theorem 4 reads as follows: $\exp'_b (x) = \exp'_b (0)\exp_b (x)$.

Let us start with the base 10 of common logarithms and try to find another base b for which $\exp'_b (0) = 1$. By definition of the logarithm,

$$b = 10^{\log_{10} b} \quad \text{(see p. 129)}$$

Therefore

$$b^x = (10^{\log_{10} b})^x = 10^{x \log_{10} b}$$

Hence

$$\exp_b (x) = \exp_{10} (x \log_{10} b)$$

Differentiate by using the chain rule:

$$\exp'_b (x) = \exp'_{10} (x \log_{10} b) \cdot \log_{10} b$$

Set $x = 0$:

$$\exp'_b (0) = \exp'_{10} (0) \cdot \log_{10} b$$

If we pick b so that

$$\exp'_{10} (0) \cdot \log_{10} b = 1 \tag{1}$$

then we will have $\exp'_b (x) = \exp_b (x)$, as desired. Solving (1) for b, we have

$$\log_{10} b = \frac{1}{\exp'_{10} (0)}$$

That is,

$$b = \exp_{10} \left[\frac{1}{\exp'_{10} (0)} \right]$$

We denote the number $\exp_{10} [1/\exp'_{10} (0)]$ by the letter e. Its numerical value is approximately 2.7182818285, and we have

$$\exp'_e (x) = \exp_e (x)$$

Although we started with the arbitrary choice of 10 as a base, it is easy to show (see Solved Exercise 15) that any initial choice of base leads to the same value for e. Since the base e is so special, we write $\exp(x)$ for $\exp_e (x) = e^x$.

Logarithms to the base e are called *natural logarithms*. We denote $\log_e x$ by $\ln x$. (The notation $\log x$ is generally used in calculus books for the common logarithm $\log_{10} x$.) Since $e^1 = e$, we have $\ln e = 1$.

Worked Example 8 Simplify $\ln(e^5) + \ln(e^{-3})$.

Solution By the laws of logarithms, $\ln(e^5) + \ln(e^{-3}) = \ln(e^5 \cdot e^{-3}) = \ln(e^2) = 2$.

We can now complete our differentiation formula for the general exponential function $\exp_b x$. Since $b = e^{\ln b}$, we have $b^x = e^{x \ln b}$. Using the chain rule, we find

$$\frac{d}{dx} b^x = \frac{d}{dx} e^{x \ln b}$$

$$= e^{x \ln b} \frac{d}{dx} (x \ln b)$$

$$= e^{x \ln b} \ln b$$

$$= b^x \ln b$$

Thus the mysterious factor $\exp'_b (0)$ turns out to be just the natural logarithm of b.

Worked Example 9 Differentiate: (a) $f(x) = e^{3x}$; (b) $g(x) = 3^x$.

Solution

(a) Let $u = 3x$ so $e^{3x} = e^u$ and use the chain rule:

$$\frac{d}{dx} e^u = \left(\frac{d}{du} e^u \right) \frac{du}{dx}$$

$$= e^u \cdot 3 = 3e^{3x}$$

(b) $\frac{d}{dx} 3^x = 3^x \ln 3$.

This expression cannot be simplified further; one can find the value $\ln 3 \approx 1.0986$ in a table or with a calculator.

Solved Exercises

14. Differentiate the following functions.

    (a) e^{2x} (b) 2^x

    (c) xe^{3x} (d) $\exp(x^2 + 2x)$

    (e) x^2

15. Show that, for any base b, $\exp_b(1/\exp_b'(0)) = e$.

16. Differentiate:

    (a) $e^{\sqrt{x}}$ (b) $e^{\sin x}$

    (c) $2^{\sin x}$ (d) $(\sin x)^2$

17. Prove that for $t > 0$ and $b > 1$, we have $b^t - 1 < (b^t \log_e b)t$.

Exercises

21. Differentiate the following functions.

    (a) e^{x^2+1} (b) $\sin(e^x)$ (c) $3^x - 2^{x-1}$

    (d) $e^{\cos x}$ (e) $\tan(3^{2x})$ (f) $e^{1-x^2} + x^3$

22. Differentiate (assume f and g differentiable where necessary):

    (a) $(x^3 + 2x - 1)e^{x^2 + \sin x}$ (b) $e^{2x} - \cos(x + e^{2x})$

    (c) $(e^{3x^3+x})(1 - e^x)$ (d) $(e^{x+1} + 1)(e^{x-1} - 1)$

    (e) $f(x) \cdot e^x + g(x)$ (f) $e^{f(x)+x^2}$

    (g) $f(x) \cdot e^{g(x)}$ (h) $f(e^x + g(x))$

23. Show that $f(x) = e^x$ is an increasing function.

24. Find the critical points of $f(x) = x^2 e^{-x}$.

25. Find the critical points of $f(x) = \sin x e^x$, $-4\pi \leqslant x \leqslant 4\pi$.

26. Simplify the following expressions:

    (a) $\ln(e^{x+1}) + \ln(e^2)$ (b) $\ln(e^{\sin x}) - \ln(e^{\cos x})$

The Derivative of the Logarithm

We can differentiate the logarithm function by using the inverse function rule of Chapter 8. If $y = \ln x$, then $x = e^y$ and

$$\frac{dy}{dx} = \frac{1}{dx/dy} = \frac{1}{e^y} = \frac{1}{x}$$

Hence

$$\frac{d}{dx} \ln x = \frac{1}{x}$$

For other bases, we use the same process:

$$\frac{d}{dx} \log_b x = \frac{1}{\frac{d}{dy} b^y} = \frac{1}{\ln b \cdot b^y} = \frac{1}{\ln b \cdot x}$$

That is,

$$\frac{d}{dx} \log_b x = \frac{1}{(\ln b)x}$$

The last formula may also be proved by using law 3 of logarithms:

$$\ln x = \log_e x = \log_b x \cdot \ln b$$

so

$$\frac{d}{dx} \log_b x = \frac{d}{dx} \left(\frac{1}{\ln b} \ln x \right) = \frac{1}{\ln b} \frac{d}{dx} \ln x = \frac{1}{(\ln b) \cdot x}$$

Worked Example 10 Differentiate: (a) $\ln(3x)$; (b) $xe^x \ln x$; (c) $8 \log_3 8x$.

Solution

(a) By the chain rule, setting $u = 3x$,

$$\frac{d}{dx} \ln 3x = \frac{d}{du}(\ln u) \cdot \frac{du}{dx} = \frac{1}{3x} \cdot 3 = \frac{1}{x}$$

Alternatively, $\ln 3x = \ln 3 + \ln x$, so the derivative with respect to x is $1/x$.

(b) By the product rule:

$$\frac{d}{dx}(xe^x \ln x) = x \frac{d}{dx}(e^x \ln x) + e^x \ln x = xe^x \ln x + e^x + e^x \ln x$$

(c) From the formula $(d/dx) \log_b x = 1/(\ln b)x$ with $b = 3$,

$$\frac{d}{dx} 8 \log_3 8x = 8 \frac{d}{dx} \log_3 8x$$

$$= 8\left(\frac{d}{du}\log_3 u\right)\frac{du}{dx} \quad (u = 8x)$$

$$= 8 \cdot \frac{1}{\ln 3 \cdot u} \cdot 8$$

$$= \frac{64}{(\ln 3)8x} = \frac{8}{(\ln 3)x}$$

In order to differentiate certain expressions it is sometimes convenient to begin by taking logarithms.

Worked Example 11   Differentiate the function $y = x^x$.

Solution   We take natural logarithms,

$$\ln y = \ln(x^x) = x \ln x$$

Next we differentiate, remembering that y is a function of x:

$$\frac{1}{y}\frac{dy}{dx} = x \cdot \frac{1}{x} + \ln x = 1 + \ln x$$

Hence

$$\frac{dy}{dx} = y(1 + \ln x) = x^x(1 + \ln x)$$

In general, $(d/dx)\ln f(x) = f'(x)/f(x)$ is called the *logarithmic derivative* of f. Other applications are given in the exercises which follow.

Solved Exercises

18. Differentiate:
    (a) $\ln 10x$ (b) $\ln u(x)$ (c) $\ln(\sin x)$
    (d) $(\sin x)\ln x$ (e) $(\ln x)/x$ (f) $\log_5 x$

19. (a) If n is any *real number*, prove that

    $$\frac{d}{dx}x^n = nx^{n-1} \quad \text{for } x > 0$$

    (b) Find $(d/dx)(x^\pi)$.

20. Use logarithmic differentiation to calculate dy/dx, if y is given by $y = (2x + 3)^{3/2}/\sqrt{x^2 + 1}$.

21. Differentiate $y = x^{(x^x)}$.

Exercises

27. Differentiate:

    (a) $\ln(2x + 1)$ (b) $\ln(x^2 - 3x)$ (c) $\ln(\tan x)$
    (d) $(\ln x)^3$ (e) $(x^2 - 2x)\ln(2x + 1)$ (f) $e^{x + \ln x}$
    (g) $[\ln(\tan 3x)]/(1 + \ln x^2)$

28. Use logarithmic differentiation to differentiate:

    (a) $y = x^{3x}$ (b) $y = x^{\sin x}$
    (c) $y = (\sin x)^{\cos x}$ (d) $y = (x^3 + 1)^{x^2 - 2}$
    (e) $y = (x - 2)^{2/3}(4x + 3)^{8/7}$

Problems for Chapter 10

1. Simplify:

    (a) $e^{4x}[\ln(e^{3x-1}) - \ln(e^{1-x})]$ (b) $e^{(x \ln 3 + \ln 2x)}$

2. Differentiate:

    (a) $e^{x \sin x}$ (b) x^e
    (c) $14^{x^2 - 8 \sin x}$ (d) x^{x^2}
    (e) $\ln(x^{-5} + x)$ (f) $(\ln x)^{\exp x}$
    (g) $\sin(x^4 + 1) \cdot \log_8(14x - \sin x)$

3. Sketch the graph of $y = xe^{-x}$; indicate on your graph the regions where y is increasing, decreasing, concave upward or downward.

4. Find the minimum of $y = x^x$ for x in $(0, \infty)$.

5. Suppose that f is continuous and that $f(x + y) = f(x)f(y)$ for all x and y. Show that $f(x) = b^x$ for some b. [*Hint*: Try showing that f is actually differentiable at 0.]

6. Let f be a twice differentiable convex function. Prove that $f''(x) \geqslant 0$.

7. Let f be an increasing continuous convex function. Let f^{-1} be the inverse function. Show that $-f^{-1}$ is convex. Apply this result to $\log_b x$.

8. Suppose that $f(x)$ is a function defined for all real x. If $x_1 < x_2 < x_3$ and $f(x_1) = f(x_2) = f(x_3)$, prove that f is not strictly convex. Give an example to show that f may be convex.

9. Suppose that $f(x)$ is a strictly convex differentiable function defined on $(-\infty, \infty)$. Show that the tangent line to the graph of $y = f(x)$ at (x_0, y_0) does not intersect the graph at any other point. Here, x_0 is any real number and $y_0 = f(x_0)$. What can we say if we only assume f to be convex?

10. Suppose that $f(x)$ is defined for all real x and that f is strictly convex on

$(-\infty, 0)$ and strictly convex on $(0, \infty)$. Prove that if f is convex on $(-\infty, \infty)$, then f is strictly convex on $(-\infty, \infty)$.

11. Prove that $e^x > 1 + x^2$, for $x \geqslant 1$. [*Hint*: Note that $e > 2$ and show that the difference between these two functions is increasing.]

12. We have seen that the exponential function $\exp(x)$ satisfies the following relations: $\exp(x) > 0$, $\exp(0) = 1$, and $\exp'(x) = \exp(x)$. Let $f(x)$ be a function such that

$$0 \leqslant f'(x) \leqslant f(x) \text{ for all } x \geqslant 0$$

Prove that $0 \leqslant f(x) \leqslant f(0) \exp(x)$ for all $x \geqslant 0$. (*Hint*: Consider $g(x) = f(x)/\exp(x)$.)

13. Let $f(x)$ be an increasing continuous function. Given x_0, let

$A =$ the set of $f(x)$ where $x < x_0$

$B =$ the set of $f(x)$ where $x > x_0$

Show that $f(x_0)$ is the transition point from A to B.

14. Show that the sum $f(x) + g(x)$ of two convex functions $f(x)$ and $g(x)$ is convex. Show that if f is strictly convex, then so is $f + g$. Use this to show that $f(x) = ax^2 + bx + c$ is strictly convex if $a > 0$, where a, b, and c are constants.

15. Suppose $f(x)$ is defined and differentiable for all real x.
    (a) Does $f''(x) \geqslant 0$ for all x imply that $f(x)$ is convex?
    (b) Does $f''(x) > 0$ for all x imply that $f(x)$ is convex? Strictly convex?

11   The Integral

In this chapter we define the integral in terms of transitions; i.e., by the method of exhaustion. The reader is assumed to be familiar with the summation notation and its basic properties, as presented in most calculus texts.

Piecewise Constant Functions

In the theory of differentiation, the simplest functions were the linear functions $f(x) = ax + b$. We knew that the derivative of $ax + b$ should be a, and we defined the derivative for more general functions by comparison with the linear functions, using the notion of overtaking to make the comparisons.

For integration theory, the comparison functions are the *piecewise constant functions*. Roughly speaking, a function f on $[a, b]$ is piecewise constant if $[a, b]$ can be broken into a finite number of subintervals such that f is constant on each subinterval.

Definition A *partition* of the interval $[a, b]$ is a sequence of numbers $(t_0, t_1, \ldots, t_n)$ such that

$$a = t_0 < t_1 < \cdots < t_{n-1} < t_n = b$$

We think of the numbers $t_0, t_1, \ldots, t_n$ as dividing $[a, b]$ into the subintervals $[t_0, t_1], [t_1, t_2], \ldots, [t_{n-1}, t_n]$. (See Fig. 11-1.) The number n of intervals can be as small as 1 or as large as we wish.

Fig. 11-1 A partition of $[a, b]$.

Definition A function f defined on an interval $[a, b]$ is called *piecewise constant* if there is a partition $(t_0, \ldots, t_n)$ of $[a, b]$ and real numbers $k_1, \ldots, k_n$ such that

$$f(t) = k_i \quad \text{for all } t \text{ in } (t_{i-1}, t_i)$$

The partition $(t_0, \ldots, t_n)$ is then said to be *adapted* to the piecewise constant function f.

Notice that we put no condition on the value of f at the points $t_0, t_1, \ldots, t_n$, but we require that f be constant on each of the *open* intervals between successive points of the partition. As we will see in Worked Example 1, more than one partition may be adapted to a given piecewise constant function. Piecewise constant functions are sometimes called *step functions* because their graphs often resemble staircases (see Fig. 11-2).

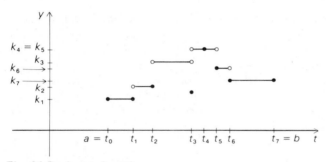

Fig. 11-2 A step function.

Worked Example 1 Draw a graph of the piecewise constant function f on $[0,1]$ defined by

$$f(t) = \begin{cases} -2 & \text{if } 0 \leqslant t < \frac{1}{3} \\ 3 & \text{if } \frac{1}{3} \leqslant t < \frac{1}{2} \\ 3 & \text{if } \frac{1}{2} \leqslant t \leqslant \frac{3}{4} \\ 1 & \text{if } \frac{3}{4} < t \leqslant 1 \end{cases}$$

Give three different partitions which are adapted to f and one partition which is not adapted to f.

Solution The graph is shown in Fig. 11-3. As usual, an open circle indicates a point which is not on the graph. (The resemblance of this graph to a staircase is rather faint.)

One partition adapted to f is $(0, \frac{1}{3}, \frac{1}{2}, \frac{3}{4}, 1)$. (That is, $t_0 = 0$, $t_1 = \frac{1}{3}$, $t_2 = \frac{1}{2}$, $t_3 = \frac{3}{4}$, $t_4 = 1$. Here $k_1 = -2, k_2 = k_3 = 3, k_4 = 1$.) If we delete $\frac{1}{2}$, the

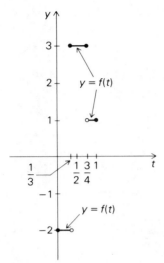

Fig. 11-3 Another step function.

partition $(0, \frac{1}{3}, \frac{3}{4}, 1)$ is still adapted to f because f is constant on $(\frac{1}{3}, \frac{3}{4})$. Finally, we can always add extra points to an adapted partition and it will still be adapted. For example, $(0, \frac{1}{8}, \frac{1}{3}, \frac{1}{2}, \frac{3}{4}, \frac{8}{9}, 1)$ is an adapted partition.

A partition which is not adapted to f is $(0, 0.1, 0.2, 0.3, 0.4, 0.5, 0.6, 0.7, 0.8, 0.9, 1)$. Even though the intervals in this partition are very short, it is not adapted to f because f is not constant on $(0.7, 0.8)$. (Can you find another open interval in the partition on which f is not constant?)

Motivated by our physical example, we define the integral of a piecewise constant function as a sum.

Definition Let f be a piecewise constant function on $[a, b]$. Let $(t_0, \ldots, t_n)$ be a partition of $[a, b]$ which is adapted to f and let k_i be the value of f on (t_{i-1}, t_i). The sum

$$k_1(t_1 - t_0) + k_2(t_2 - t_1) + \cdots + k_n(t_n - t_{n-1})$$

is called the *integral* of f on $[a, b]$ and is denoted by $\int_a^b f(t)\,dt$; that is,

$$\int_a^b f(t)\,dt = \sum_{i=1}^n k_i \Delta t_i$$

where $\Delta t_i = t_i - t_{i-1}$.

We shall verify shortly that the definition is independent of the choice of adapted partition.

If all the k_i's are nonnegative, the integral $\int_a^b f(t)\,dt$ is precisely the area "under" the graph of f—that is, the area of the set of points (x, y) such that $a \leqslant x \leqslant b$ and $0 \leqslant y \leqslant f(x)$. The region under the graph in Fig. 11-2 is shaded in Fig. 11-4.

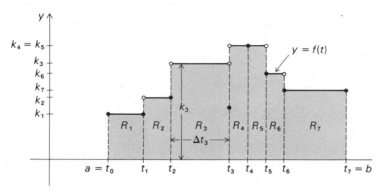

Fig. 11-4 The shaded area is the sum of the areas of the rectangles R_i.

The notation $\int_a^b f(t)\,dt$ for the integral is due to Leibniz. The symbol $\int$, called an *integral sign*, is an elongated S which replaces the Greek Σ of ordinary summation. Similarly, the dt replaces the Δt_i's in the summation formula. The function $f(t)$ which is being integrated is called the *integrand*. The endpoints a and b are also called *limits of integration*.

Worked Example 2 Compute $\int_0^1 f(t)\,dt$ for the function in Worked Example 1, first using the partition $(0, \frac{1}{3}, \frac{1}{2}, \frac{3}{4}, 1)$ and then using the partition $(0, \frac{1}{3}, \frac{3}{4}, 1)$.

Solution With the first partition, we have

$$k_1 = -2 \quad \Delta t_1 = t_1 - t_0 = \tfrac{1}{3} - 0 = \tfrac{1}{3}$$
$$k_2 = 3 \quad \Delta t_2 = t_2 - t_1 = \tfrac{1}{2} - \tfrac{1}{3} = \tfrac{1}{6}$$
$$k_3 = 3 \quad \Delta t_3 = t_3 - t_2 = \tfrac{3}{4} - \tfrac{1}{2} = \tfrac{1}{4}$$
$$k_4 = 1 \quad \Delta t_4 = t_4 - t_3 = 1 - \tfrac{3}{4} = \tfrac{1}{4}$$

Thus,

$$\int_0^1 f(t)\,dt = \sum_{i=1}^4 k_i \Delta t_i = (-2)(\tfrac{1}{3}) + (3)(\tfrac{1}{6}) + (3)(\tfrac{1}{4}) + (1)(\tfrac{1}{4})$$

$$= -\tfrac{2}{3} + \tfrac{1}{2} + \tfrac{3}{4} + \tfrac{1}{4} = \tfrac{5}{6}$$

Using the second partition, we have

$$k_1 = -2 \quad \Delta t_1 = t_1 - t_0 = \tfrac{1}{3} - 0 = \tfrac{1}{3}$$
$$k_2 = 3 \quad \Delta t_2 = t_2 - t_1 = \tfrac{3}{4} - \tfrac{1}{3} = \tfrac{5}{12}$$
$$k_3 = 1 \quad \Delta t_3 = t_3 - t_2 = 1 - \tfrac{3}{4} = \tfrac{1}{4}$$

and

$$\int_0^1 f(t)\,dt = \sum_{i=1}^3 k_i \Delta t_i = (-2)(\tfrac{1}{3}) + (3)(\tfrac{5}{12}) + (1)(\tfrac{1}{4})$$

$$= -\tfrac{2}{3} + \tfrac{5}{4} + \tfrac{1}{4} = \tfrac{5}{6}$$

which is the same answer we obtained from the first partition.

Theorem 1 *Let f be a piecewise constant function on $[a, b]$. Suppose that $(t_0, t_1, \ldots, t_n)$ and $(s_0, s_1, \ldots, s_m)$ are adapted partitions, with $f(t) = k_i$ on (t_{i-1}, t_i) and $f(t) = j_i$ on (s_{i-1}, s_i). Then*

$$\sum_{i=1}^n k_i \Delta t_i = \sum_{i=1}^m j_i \Delta s_i$$

Proof The idea is to reduce the problem to the simplest case, where the second partition is obtained from the first by the addition of a single point. Let us begin by proving the proposition for this case. Assume, then, that $m = n + 1$, and that $(s_0, s_1, \ldots, s_m) = (t_0, t_1, \ldots, t_l, t_*, t_{l+1}, \ldots, t_n)$; i.e., the s-partition is obtained from the t-partition by inserting an extra point t^* between t_l and t_{l+1}. The relation between s's, t's, j's, and k's is illustrated in Fig. 11-5. The sum obtained from the s-partition is

$$\sum_{i=1}^m j_i \Delta s_i = j_1(s_1 - s_0) + \cdots + j_{l+1}(s_{l+1} - s_l)$$

$$+ j_{l+2}(s_{l+2} - s_{l+1}) + \cdots + j_{n+1}(s_{n+1} - s_n)$$

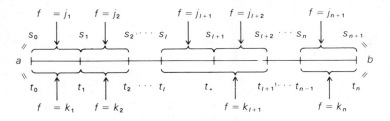

Fig. 11-5 The s-partition is finer than the t-partition.

Substituting the appropriate k's and t's for the j's and s's converts the sum to

$$k_1(t_1 - t_0) + \cdots + k_{l+1}(t_* - t_l) + k_{l+1}(t_{l+1} - t_*) + \cdots + k_n(t_n - t_{n-1})$$

We can combine the two middle terms:

$$k_{l+1}(t_* - t_l) + k_{l+1}(t_{l+1} - t_*) = k_{l+1}(t_* - t_l + t_{l+1} - t_*)$$
$$= k_{l+1}(t_{l+1} - t_l)$$

The sum is now

$$k_1(t_1 - t_0) + \cdots + k_{l+1}(t_{l+1} - t_l) + \cdots + k_n(t_n - t_{n-1}) = \sum_{i=1}^{n} k_i \Delta t_i$$

which is the sum obtained from the t-partition. (For a numerical illustration, see Worked Example 2.) This completes the proof for the special case.

To handle the general case, we observe first that, given two partitions $(t_0, \ldots, t_n)$ and $(s_0, \ldots, s_m)$, we can find a partition $(u_0, \ldots, u_p)$ which contains both of them taking all the points $t_0, \ldots, t_n, s_0, \ldots, s_m$, eliminating duplications, and putting the points in the correct order. (See Solved Exercise 1, immediately following the end of the proof.) Adding points to an adapted partition produces another adapted partition, since if a function is constant on an interval, it is certainly constant on any subinterval. It follows that the u-partition is adapted to f if the s- and t-partitions are. Now we can get from the t-partition to the u-partition by adding points one at a time. By the special case above, we see that the sum is unchanged each time we add a point, so the sum obtained from the u-partition equals the sum obtained from the t-partition. In a similar way, we can get from the s-partition to the u-partition by adding one point at a time, so the sum from the u-partition equals the sum from the s-partition. Since the sums from the t- and s-partitions are both equal to the sum from the u-partition, they are equal to each other, which is what we wanted to prove.

Solved Exercises

1. Consider s- and t-partitions of $[1, 8]$ as follows. Let the s-partition be $(1, 2, 3, 4, 7, 8)$, and let the t-partition be $(1, 4, 5, 6, 8)$. Find the corresponding u-partition, and show that you can get from the s- and t-partitions to the u-partition by adding one point at a time.

2. Let $f(t)$ be defined by

$$f(t) = \begin{cases} 2 & \text{if } 0 \leqslant t < 1 \\ 0 & \text{if } 1 \leqslant t < 3 \\ -1 & \text{if } 3 \leqslant t \leqslant 4 \end{cases}$$

For any number x in $(0,4]$, $f(t)$ is piecewise constant on $[0,x]$.

(a) Find $\int_0^x f(t)\,dt$ as a function of x. (You will need to use different formulas on different intervals.)

(b) Let $F(x) = \int_0^x f(t)dt$, for $x \in (0,4]$. Draw a graph of F.

(c) At which points is F differentiable? Find a formula for $F'(x)$.

Exercises

1. Let $f(t)$ be the greatest integer function: $f(t) = n$ on $(n, n+1)$, where n is any integer. Compute $\int_0^8 f(t)dt$ using each of the partitions $(1,2,3,4,5,6,7,8)$ and $(1,2,2.5,3,3.5,4,5,6,7,8)$. Sketch.

2. Show that, given any two piecewise constant functions on the same interval, there is a partition which is adapted to both of them.

Upper and Lower Sums and the Definition

Having defined the integral for piecewise constant functions, we will now define the integral for more general functions. You should compare the definition with the way in which we passed from linear functions to general functions when we defined the derivative in Chapters 1 and 2. We begin with a preliminary definition.

Definition   Let f be a function defined on $[a, b]$. If g is any piecewise constant function on $[a, b]$ such that $g(t) \leqslant f(t)$ for all t in the open interval (a, b), we call the number $\int_a^b g(t)dt$ a *lower sum* for f on $[a, b]$. If h is a piecewise constant function and $f(t) \leqslant h(t)$ for all t in (a, b), the number $\int_a^b h(t)\,dt$ is called an *upper sum* for f on $[a, b]$.

Worked Example 3 Let $f(t) = t^2 + 1$ for $0 \leqslant t \leqslant 2$. Let

$$g(t) = \begin{cases} 0 & 0 \leqslant t \leqslant 1 \\ 2 & 1 < t \leqslant 2 \end{cases} \quad \text{and} \quad h(t) = \begin{cases} 2 & 0 \leqslant t \leqslant \frac{2}{3} \\ 4 & \frac{2}{3} < t \leqslant \frac{4}{3} \\ 5 & \frac{4}{3} < t \leqslant 2 \end{cases}$$

Draw a graph showing $f(t), g(t)$, and $h(t)$. What upper and lower sums for f can be obtained from g and h?

Solution The graph is shown in Fig. 11-6.

Since $g(t) \leqslant f(t)$ for all t in the open interval $(0, 2)$ (the graph of g lies below that of f), we have a lower sum

$$\int_0^2 g(t)\,dt = 0 \cdot 1 + 2 \cdot 1 = 2$$

Since the graph of h lies above that of f, $h(t) \geqslant f(t)$ for all t in the interval $(0, 2)$, so we have the upper sum

$$\int_0^2 h(t)\,dt = 2 \cdot \tfrac{2}{3} + 4 \cdot \tfrac{2}{3} + 5 \cdot \tfrac{2}{3} = \tfrac{22}{3} = 7\tfrac{1}{3}$$

The integral of a function should lie between the lower sums and the upper sums. For instance, the integral of the function in Worked Example 3 should lie in the interval $[2, 7\tfrac{1}{3}]$. If we could find upper and lower sums which are arbitrarily close together, then the integral would be pinned down to a single point. This idea leads to the formal definition of the integral.

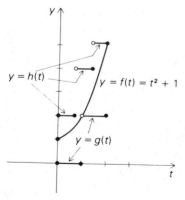

Fig. 11-6 $g(t) \leqslant f(t) \leqslant h(t)$.

Definition   Let f be a function defined on $[a, b]$, and let L_f and U_f be the sets of lower and upper sums for f, as defined above. If there is a transition point from L_f to U_f, then we say that f is *integrable* on $[a, b]$. The transition point is called *the integral of f on* $[a, b]$ and is denoted by

$$\int_a^b f(t)\, dt$$

Remark   The letter t is called the *variable of integration*; it is a dummy variable in that we can replace it by any other letter without changing the value of the integral; a and b are called the *endpoints* for the integral.

The next theorem contains some important facts about upper and lower sums.

Theorem 2

1. *Every lower sum for f is less than or equal to every upper sum.*

2. *Every number less than a lower sum is a lower sum; every number greater than an upper sum is an upper sum.*

Proof   To prove part 1 we must show that, if g and h are piecewise constant functions on $[a, b]$ such that $g(t) \leqslant f(t) \leqslant h(t)$ for all t in (a, b), then

$$\int_a^b g(t)\, dt \leqslant \int_a^b h(t)\, dt$$

The function f will play no role in this proof; all we use is the fact that $g(t) \leqslant h(t)$ for all $t \in (a, b)$.

By Theorem 1 we can use any adapted partition we want to compute the integrals of g and h. In particular, we can combine partitions which are adapted to g and h to obtain a partition which is adapted to both g and h. See Exercise 2. Let $(t_0, t_1, \ldots, t_n)$ be such a partition. Then we have constants k_i and l_i such that $g(t) = k_i$ and $h(t) = l_i$ for t in (t_{i-1}, t_i). By hypothesis, we have $k_i \leqslant l_i$ for each i. (See Fig. 11-7.) Now, each $\Delta t_i = t_i - t_{i-1}$ is positive, so $k_i \Delta t_i \leqslant l_i \Delta t_i$ for each i. Therefore,

$$\sum_{i=1}^n k_i \Delta t_i \leqslant \sum_{i=1}^n l_i \Delta t_i$$

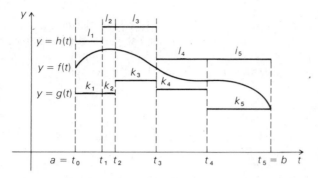

Fig. 11-7 Upper sums are larger than lower sums.

But these sums are just the integrals of g and h, so we are done with part 1.

We will now prove the first half of part 2, that every number lower than a lower sum is a lower sum. (The other half is virtually identical; we leave it to the reader as an exercise.) Let β be a lower sum and let $c < \beta$ be any number. β is the integral $\int_a^b g(t)\,dt$ of a piecewise constant function g with $g(t) \leqslant f(t)$ for all $t \in (a, b)$; we wish to show that c is such an integral as well. To do this, we choose an adapted partition $(t_0, t_1, \ldots, t_n)$ for g and create a new function e by lowering g on the interval (t_0, t_1). Specifically, we let $e(t) = g(t)$ for all t not in (t_0, t_1), and we put $e(t) = g(t) - p$ for t in (t_0, t_1), where p is a positive constant to be chosen in such a way that the integral comes out right. Specifically, if $g(t) = k_i$ for t in (t_{i-1}, t_i), then $e(t) = k_1 - p$ for t in (t_0, t_1), and $e(t) = k_i$ for t in $(t_{i-1}, t_i), i \geqslant 2$. We have

$$\int_a^b e(t)\,dt = (k_1 - p)\,\Delta t_1 + k_2\,\Delta t_2 + \cdots + k_n\,\Delta t_n$$

$$= k_1\,\Delta t_1 + \cdots + k_n\,\Delta t_n - p\,\Delta t_1$$

$$= \int_a^b g(t)\,dt - p\,\Delta t_1$$

Setting this equal to c and solving for p gives

$$p = \frac{1}{\Delta t_1}\left[\int_a^b g(t)\,dt - c\right]$$

which is positive, since we assumed $c < \beta = \int_a^b g(t)\,dt$. With this value of p, we define e as we just described (it is important that p be positive, so that $e(t) \leqslant f(t)$ for all t in (a, b)), and so we have $\int_a^b e(t)\,dt = c$, as desired.

It follows from the completeness axiom in Chapter 4 that L_f and U_f are intervals which, if nonempty, are infinite in opposite directions.

In Fig. 11-8 we show the possible configurations for L_f and U_f. If there is a transition point from L_f to U_f, there is exactly one, so the integral, if it exists, is well defined.

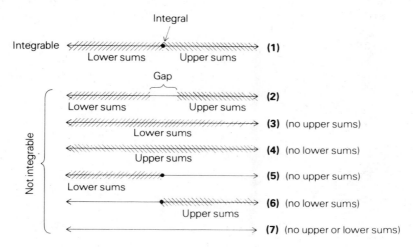

Fig. 11-8 Possible configurations of lower and upper sums.

Do not conclude from Fig. 11-8 that most functions are nonintegrable. In fact, cases (3), (4), (5), (6), and (7) can occur only when f is unbounded (see Worked Example 4). The functions for which L_f and U_f has a gap between them (case (2)) are quite "pathological" (see Solved Exercise 3). In fact, Theorem 3 in the next section tells us that integrability is even more common than differentiability.

Worked Example 4 Show that the set of upper sums for f is nonempty if and only if f is bounded above on $[a, b]$; i.e., if and only if there is a number M such that $f(t) \leqslant M$ for all t in $[a, b]$.

Solution Suppose first that $f(t) \leqslant M$ for all t in $[a, b]$. Then we can consider the piecewise constant function g defined by $g(t) = M$ for all t in $[a, b]$. We have $f(t) \leqslant g(t)$ for all t in $[a, b]$, so

$$\int_a^b h(t)\,dt = M(b-a)$$

is an upper sum for f.

The converse is a little more difficult. Suppose that there exists an upper sum for f. This upper sum is the integral of a piecewise constant function h on $[a, b]$ such that $f(t) \leqslant h(t)$ for all t in $[a, b]$. It suffices to show that the piecewise constant function h is bounded above on $[a, b]$. To do this, we choose a partition $(t_0, t_1, \ldots, t_n)$ for h. For every t in $[a, b]$, the value of $h(t)$ belongs to the finite list

$$h(t_0), k_1, h(t_1), k_2, \ldots, h(t_{n-1}), k_n, h(t_n)$$

where k_i is the value of h on (t_{i-1}, t_i). If we let M be the largest number in the list above, we may conclude that $h(t)$, and hence $f(t)$, is less than or equal to M for all $t \in [a, b]$. (See Fig. 11-9.)

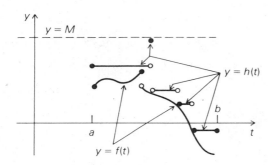

Fig. 11-9 The piecewise constant function h is bounded above by M on $[a, b]$, so f is bounded above too.

Solved Exercises

3. Let

$$f(t) = \begin{cases} 0 & \text{if } t \text{ is irrational} \\ 1 & \text{if } t \text{ is rational} \end{cases}$$

for $0 \leqslant t \leqslant 1$. Show that f is not integrable.

4. Show that every piecewise constant function is integrable and that its integral as defined on p. 155 is the same as its integral as originally defined on p. 149.

Exercises

3. Let f be defined on $[0, 1]$ by

$$f(t) = \begin{cases} 1 & \text{if } t \text{ is rational} \\ 0 & \text{if } t \text{ is irrational} \end{cases}$$

Find the sets of upper and lower sums for f and show that f is not integrable.

4. Let f be a function defined on $[a, b]$ and let S_0 be a real number. Show that S_0 is *the integral of f on* $[a, b]$ if and only if:

    1. Every number $S < S_0$ is a lower sum for f on $[a, b]$.

    2. Every number $S > S_0$ is an upper sum for f on $[a, b]$.

The Integrability of Continuous Functions

We will now prove that a continuous function on a closed interval is integrable. Since every differentiable function is continuous, we conclude that every differentiable function is integrable; however, some noncontinuous functions can be integrable. For instance, piecewise constant functions are integrable (see Solved Exercise 4), but they are not continuous. On the other hand, the wild function described in Solved Exercise 3 shows that there is really something to prove. Since the proof is similar in spirit to those in Chapter 5, it may be useful to review that section before proceeding.

Theorem 3 *If f is continuous on $[a, b]$, then f is integrable on $[a, b]$.*

Proof Since the continuous function f is bounded, it has both upper and lower sums. To show that there is no gap between the upper and lower sums, we will prove that, for any positive number ϵ, there are lower and upper sums within ϵ of one another. (The result of Problem 13, Chapter 4 implies that f is integrable.) To facilitate the proof, if $[a, x]$ is any sub-interval of $[a, b]$, and ϵ is a positive real number, we will say that f is ϵ-*integrable* on $[a, x]$ if there are piecewise constant functions g and h on $[a, x]$ with $g(t) < f(t) < h(t)$ for all t in (a, x), such that $\int_a^x h(t)\,dt - \int_a^x g(t)\,dt < \epsilon$. What we wish to show, then, is that the continuous function f is ϵ-integrable on $[a, b]$, for any positive ϵ.

We define the set S_ϵ (for any $\epsilon > 0$) to consist of those x in $(a, b]$ for which f is ϵ-integrable on $[a, x]$. We wish to show that $b \in S_\epsilon$.

We will use the completeness axiom, so we begin by showing that S_ϵ is convex. If $x_1 < x < x_2$, and x_1 and x_2 are in S_ϵ, then x is certainly in $(a, b]$. To show that f is ϵ-integrable on $(a, x]$, we take the piecewise constant functions which do the job on $[a, x_2]$ (since $x_2 \in S_\epsilon$) and restrict them to $[a, x]$. The completeness axiom implies that S_ϵ is an interval. We analyze the interval S_ϵ in two lemmas, after which we will end the proof of Theorem 3.

Lemma 1 S_ϵ *contains all x in (a, c) for some $c > a$.*

Proof We use the continuity of f at a. Let $\delta = \epsilon/[2(b-a)]$. Since $f(a) - \delta < f(a) < f(a) + \delta$, there is an interval $[a, c)$ such that $f(a) - \delta < f(t) < f(a) + \delta$ for all t in $[a, c)$. Now for x in $[a, c]$, if we restrict f to $[a, x]$, we can contain it between the constant function $g(t) = f(a) - \delta$ and $h(t) = f(a) + \delta$. Therefore,

$$\int_a^x h(t)\,dt - \int_a^x g(t)\,dt = (f(a) + \delta)(x - a) - (f(a) - \delta)(x - a)$$

$$= (f(a) + \delta - f(a) + \delta)(x - a)$$

$$= 2\delta(x - a) = 2 \cdot \frac{\epsilon}{2(b-a)}(x - a)$$

$$= \epsilon\, \frac{x - a}{b - a} < \epsilon \quad \text{since } x < c \leqslant b,$$

and we have shown that f is ϵ-integrable on $[a, x]$. Therefore, x belongs to S_ϵ as required.

Lemma 2 *If $x_0 < a$, and x_0 belongs to S_ϵ, then S_ϵ contains all x in $(x_0, c]$ for some $c > x_0$.*

Proof This is a slightly more complicated variation of the previous proof. By the hypothesis concerning x_0, there are piecewise constant functions g_0 and h_0 on $[a, x_0]$ with $g_0(t) \leqslant f(t) \leqslant h_0(t)$ for all t in (a, x_0), such that the difference $\delta = \int_a^{x_0} h_0(t)\,dt - \int_a^{x_0} g_0(t)\,dt$ is less than ϵ. By the continuity of f at x_0 there is a number $c > x_0$ such that

$$f(x_0) - \frac{\epsilon - \delta}{2(b-a)} < f(t) < f(x_0) + \frac{\epsilon - \delta}{2(b-a)}$$

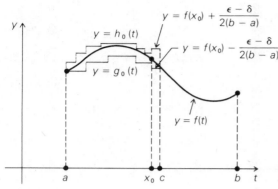

Fig. 11-10 Extending g_0 and h_0 from $[a, x_0]$ to $[a, c]$.

for all t in $[x_0, c)$.

Now we extend the functions g_0 and h_0 to the interval $[a, c]$ by defining g and h as follows (Fig. 11-10):

$$g(t) = \begin{cases} g_0(t) & a \leqslant t \leqslant x_0 \\ \\ f(x_0) - \dfrac{\epsilon - \delta}{2(b - a)} & x_0 < t \leqslant c \end{cases}$$

$$h(t) = \begin{cases} h_0(t) & a \leqslant t \leqslant x_0 \\ \\ f(x_0) + \dfrac{\epsilon - \delta}{2(b - a)} & x_0 < t < c \end{cases}$$

Clearly, we have $g(t) \leqslant f(t) \leqslant h(t)$ for all t in $[a, c]$, We claim now that $\int_a^c h(t)\, dt - \int_a^c g(t)\, dt < \epsilon$. To show this, we observe first that

$$\int_a^c g(t)\, dt = \int_a^{x_0} g(t)\, dt + \left[f(x_0) - \frac{\epsilon - \delta}{2(b - a)} \right](c - x_0)$$

(In fact, the sum for $\int_a^c g(t)\, dt$ is just that for $\int_a^{x_0} g(t)\, dt$ with one more term added for the interval $[x_0, c]$.) Similarly, $\int_a^c h(t)\, dt = \int_a^{x_0} h(t)\, dt + [f(x_0) + (\epsilon - \delta)/2(b - a)](c - x_0)$. Subtracting the last two integrals gives

$$\int_a^c h(t)\, dt - \int_a^c g(t)\, dt = \int_a^{x_0} h(t)\, dt - \int_a^{x_0} g(t)\, dt$$

$$+ [f(x_0) + \frac{\epsilon - \delta}{2(b-a)}](c - x_0) - [f(x_0) - \frac{\epsilon - \delta}{2(b-a)}](c - x_0)$$

$$= \int_a^{x_0} h(t)\,dt - \int_a^{x_0} g(t)\,dt + \frac{(\epsilon - \delta)(c - x_0)}{(b-a)}$$

Since $(c - x_0)/(b - a) < 1$, the last expression is less than $\int_a^{x_0} h(t)\,dt - \int_a^{x_0} g(t)\,dt + \epsilon - \delta = \delta + \epsilon - \delta = \epsilon$, and we have shown that f is ϵ-integrable on $[a, c]$, so c belongs to S_ϵ. Since S_ϵ is an interval, S_ϵ contains all of $(a, c]$.

Proof of Theorem 3 (completed)   By Lemma 1, S_ϵ is nonempty for every ϵ. By Lemma 2, the right-hand endpoint of S_ϵ cannot be less than b, so $S_\epsilon = [a, b)$ or $[a, b]$ for every $\epsilon > 0$. To show that S_ϵ is actually $[a, b]$ and not $[a, b)$ we use once more an argument like that in the lemmas. By continuity of f at b, we can find a number $c < b$ such that $f(b) - \epsilon/2(b - a) < f(t) < f(b) + \epsilon/2(b - a)$ for all t in $[c, b]$. Now we use the fact that $c \in S_{\epsilon/2}$, i.e., f is $\epsilon/2$-integrable on $[a, c]$. (Remember, we have established that f is ϵ-integrable on $[a, c]$ for *all* numbers ϵ, so we can use $\epsilon/2$ for ϵ.) As we did in Lemma 2, we can put together piecewise constant functions g_0 and h_0 for f on $[a, c]$ with the constant functions $f(b) \pm \epsilon/2(b - a)$ on $(c, b]$ to establish that f is ϵ-integrable on $[a, b]$, i.e., that $S_\epsilon = [a, b]$.

We have shown that f has lower and upper sums on $[a, b]$ which are arbitrarily close together, so there is no gap between the intervals L_f and U_f and f is integrable on $[a, b]$.

Solved Exercises

5. Show that, if f is ϵ-integrable on $[a, x_0]$, and $a < x < x_0$, then f is ϵ-integrable on $[a, x]$.

6. Show that every monotonic function is integrable.

7. Find $\int_1^2 (1/t)\,dt$ to within an error of no more than $\frac{1}{10}$.

8. If you used a method analogous to that in Solved Exercise 7, how many steps would it take to calculate $\int_1^2 (1/t)\,dt$ to within $\frac{1}{100}$?

Exercises

5. Let f be the nonintegrable function of Solved Exercise 3.

   (a) For which values of ϵ is f ϵ-integrable on $[0, 1]$?

   (b) Let x belong to $[0, 1]$. For which values of ϵ is f ϵ-integrable on $[0, x]$?

6. Prove that if g is piecewise constant on $[a, c]$, and $b \in (a, c)$, then

$$\int_a^c g(t) \, dt = \int_a^b g(t) \, dt + \int_b^c g(t) \, dt$$

[*Hint:* Choose a partition which includes c as one of its points.]

7. Prove that if g_1 and g_2 are piecewise constant on $[a, b]$, and s_1 and s_2 are constants, then $s_1 g_1 + s_2 g_2$ is also piecewise constant, and

$$\int_a^b (s_1 g_1(t) + s_2 g_2(t)) \, dt = s_1 \int_a^b g_1(t) \, dt + s_2 \int_a^b g_2(t) \, dt$$

[*Hint:* Choose a partition adapted to both g_1 and g_2, and write out all the sums.]

8. Let f be defined on $[0, 1]$ by

$$f(t) = \begin{cases} 0 & \text{if } t = 0 \\ \dfrac{1}{t} & \text{if } 0 < t \leqslant 1 \end{cases}$$

Is f integrable on $[0, 1]$? On $[\frac{1}{8}, 1]$?

Properties of the Integral

We now establish some basic properties of the integral. These properties imply, for example, that the integral of a piecewise constant function must be defined as we did.

Theorem 4

    1. If $a < b < c$ and f is integrable on $[a, b]$ and $[b, c]$, then f is integrable on $[a, c]$ and

$$\int_a^c f(t) \, dt = \int_a^b f(t) \, dt + \int_b^c f(t) \, dt$$

2. *If f and g are integrable on* $[a, b]$ *and if* $f(t) \leqslant h(t)$ *for all* $t \in$ (a, b), *then*

$$\int_a^b f(t)\, dt \leqslant \int_a^b h(t)\, dt$$

3. *If* $f(t) = k$ *for all* $t \in (a, b)$, *then*

$$\int_a^b f(t)\, dt = k(b - a)$$

Proof We want to show that the sum

$$\int_a^b f(t)\, dt + \int_b^c f(t)\, dt = I$$

is a transition point between the lower and upper sums for f on $[a, b]$. Let $m < I$; we will show that m is a lower sum for f on $[a, b]$. Since $m < I$, we can write $m = m_1 + m_2$, where $m_1 < \int_a^b f(t)\, dt$ and $m_2 < \int_b^c f(t)\, dt$. (See Solved Exercise 9.) Thus m_1 and m_2 are lower sums for f on $[a, b]$ and $[b, c]$, respectively. Thus, there is a piecewise constant function g_1 on $[a, b]$ with $g_1(t) < f(t)$ for all t in (a, b), such that $\int_a^b g_1(t)\, dt = m_1$, and there is a piecewise constant function g_2 on $[a, b]$ with $g_2(t) < f(t)$ for all t in (b, c) such that $\int_b^c g_2(t) = m_2$. Put together g_1 and g_2 to obtain a function g on $[a, c]$ by the definition

$$g(t) = \begin{cases} g_1(t) & a \leqslant t < b \\ f(b) - 1 & t = b \\ g_2(t) & b < t \leqslant c \end{cases}$$

The function g is piecewise constant on $[a, c]$, with $g(t) < f(t)$ for all t in (a, c). The sum which represents the integral for g on $[a, c]$ is the sum of the sums representing the integrals of g_1 and g_2, so we have

$$\int_a^c g(t)\, dt = \int_a^b g_1(t)\, dt + \int_b^c g_2(t)\, dt = m_1 + m_2 = m$$

and m is a lower sum. Similarly, any number M greater than I is an upper sum, so I is the integral of f on $[a, b]$.

We leave the proof of part 2 for the reader (Exercise 9). Part 3 follows from Solved Exercise 4.

Solved Exercises

9. If $m < p_1 + p_2$, prove that $m = m_1 + m_2$ for some numbers $m_1 < p_1$ and $m_2 < p_2$.
10. Let

$$f(x) = \begin{cases} 0 & 0 \leqslant x \leqslant 1 \\ x^2 & 1 < x \leqslant 2 \end{cases}$$

Prove that f is integrable on $[0, 2]$.

Exercises

9. Prove that, if f and g are integrable on $[a, b]$, and if $f(t) \leqslant h(t)$ for all t in (a, b), then $\int_a^b f(t)\, dt \leqslant \int_a^b h(t)\, dt$. [*Hint:* Relate the lower and upper sums for f and h.]
10. Let m, p_1, and p_2 be positive numbers such that $m < p_1 p_2$. Prove that m can be written as a product $m_1 m_2$, where $0 < m_1 < p_1$ and $0 < m_2 < p_2$.

Calculating Integrals by Hand

We have just given an elaborate definition of the integral and proved that continuous functions are integrable, but we have not yet computed the integral of any functions except piecewise constant functions. You may recall that in our treatment of differentiation it was quite difficult to compute derivatives by using the definitions; instead, we used the algebraic rules of calculus to compute most derivatives. The situation is much the same for integration. In Chapter 12 we will develop the machinery which makes the calculation of many integrals quite simple. Before doing this, however, we will calculate one integral "by hand" in order to illustrate the definition.

After the constant functions, the simplest continuous function is $f(t) = t$. We know by Theorem 1 that the integral $\int_a^b t\, dt$ exists; to calculate this integral, we will find upper and lower sums which are closer and closer together, reducing to zero the possible error in the estimate of the integral.

Let $f(t) = t$. We divide the interval $[a, b]$ into n equal parts, using the partition

$$(t_0, \ldots, t_n) = \left(a, a + \frac{b-a}{n}, a + \frac{2(b-a)}{n}, \ldots, a + \frac{(n-1)(b-a)}{n}, b \right)$$

Note that $\Delta t_i = (b-a)/n$ for each i.

Now we define the piecewise constant function g on $[a, b]$ by setting $g(t) = t_{i-1}$ for $t_{i-1} \leqslant t < t_i$. (See Fig. 11-11.)

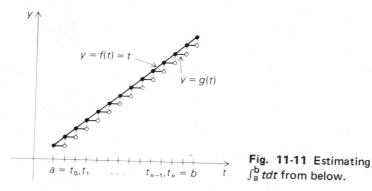

Fig. 11-11 Estimating $\int_a^b t\, dt$ from below.

Note that for $t_{i-1} \leqslant t < t_i$, we have $g(t) = t_{i-1} \leqslant t = f(t)$, so $\int_a^b g(t)\, dt$ is a lower sum for f on $[a, b]$. The definition of the integral of a piecewise constant function gives

$$\int_a^b g(t)\, dt = \sum_{i=1}^n k_i \Delta t_i = \sum_{i=1}^n t_{i-1} \Delta t_i$$

since t_{i-1} is the constant value of g on (t_{i-1}, t_i). We know that $\Delta t_i = (b-a)/n$ for each i. To find out what t_i is, we note that $t_0 = a$, $t_1 = a + (b-a)/n$, $t_2 = a + [2(b-a)/n]$, and so on so the general formula is $t_i = a + [i(b-a)/n]$. Substituting for t_{i-1} and Δt_i gives

$$\int_a^b g(t)\, dt = \sum_{i=1}^n \left[a + \frac{(i-1)(b-a)}{n} \right] \left(\frac{b-a}{n} \right)$$

$$= \sum_{i=1}^n \left[\frac{a(b-a)}{n} + \frac{i(b-a)^2}{n^2} - \frac{(b-a)^2}{n^2} \right]$$

We may rewrite this as:

$$\sum_{i=1}^{n} \frac{a(b-a)}{n} + \frac{(b-a)^2}{n^2} \sum_{i=1}^{n} i - \sum_{i=1}^{n} \frac{(b-a)^2}{n^2}$$

The outer terms do not involve i, so each may be summed by adding the summand to itself n times, i.e., by multiplying the summand by n. The middle term is summed by the formula $\sum_{i=1}^{n} i = n(n+1)/2$. The result is

$$a(b-a) + \frac{(b-a)^2}{n^2} \frac{n(n+1)}{2} - \frac{(b-a)^2}{n}$$

which simplifies to

$$\frac{b^2 - a^2}{2} - \frac{1}{2n}(b-a)^2$$

(You should carry out the simplification.) We have, therefore, shown that $[(b^2 - a^2)/2] - (1/2n)(b-a)^2$ is a lower sum for f.

We now move on to the upper sums. Using the same partition as before, but with the function $h(t)$ defined by $h(t) = t_i$ for $t_{i-1} \leqslant t < t_i$, $\int_a^b h(t)\,dt$ is an upper sum. Some algebra (see Solved Exercise 11) gives

$$\int_a^b h(t)\,dt = \frac{b^2 - a^2}{2} + \frac{(b-a)^2}{2n} \qquad \bullet$$

Our calculations of upper and lower sums therefore show that

$$\frac{b^2 - a^2}{2} - \frac{(b-a)^2}{2n} \leqslant \int_a^b t\,dt \leqslant \frac{b^2 - a^2}{2} + \frac{(b-a)^2}{2n}$$

These inequalities, which hold for all n, show that the integral can be neither larger nor smaller than $(b^2 - a^2)/2$, so it must be equal to $(b^2 - a^2)/2$.

Worked Example 5 Find $\int_0^3 x\,dx$.

Solution We first note that since the variable of integration is a "dummy variable,"

$$\int_0^3 x \, dx = \int_0^3 t \, dt$$

Using the formula just obtained, we can evaluate this integral as $\frac{1}{2}(3^2 - 0^2) = \frac{9}{2}$.

In the next chapter you will find, to your relief, a much easier way to compute integrals. Evaluating integrals like $\int_a^b t^2 \, dt$ or $\int_a^b t^3 \, dt$ by hand is possible but rather tedious. The methods of the next chapter will make these integrals simple to evaluate.

Solved Exercises

11. Draw a graph of the function $h(t)$ used above to find an upper sum for $f(t) = t$ and show that

$$\int_a^b h(t) \, dt = \frac{b^2 - a^2}{2} + \frac{(b-a)^2}{2n}$$

12. Using the definition of the integral, find $\int_a^b 5t \, dt$.

13. (a) Sketch the region under the graph of $f(t) = t$ on $[a, b]$, if $0 < a < b$.
    (b) Compare the area of this region with $\int_a^b t \, dt$.

Exercises

11. Using the definition of the integral, find a formula for $\int_a^b (t + 3) \, dt$.

12. Using the definition of the integral, find a formula for $\int_a^b (-t) \, dt$.

13. Explain the relation between $\int_{-2}^1 t \, dt$ and an area.

Problems for Chapter 11 ▬▬▬▬▬▬▬▬▬▬▬▬▬▬▬▬▬▬▬

1. Let f be the function defined by

$$f(t) = \begin{cases} 2 & 1 \leqslant t < 4 \\ 5 & 4 \leqslant t < 7 \\ 1 & 7 \leqslant t \leqslant 10 \end{cases}$$

(a) Find $\int_1^{10} f(t) \, dt$.
(b) Find $\int_2^9 f(t) \, dt$.
(c) Suppose that g is a function on $[1, 10]$ such that $g(t) \leqslant f(t)$ for all t in $[1, 10]$. What inequality can you derive for $\int_1^{10} g(t) \, dt$?

(d) With $g(t)$ as in part (c), what inequalities can you obtain for $\int_1^{10} 2g(t)\,dt$ and $\int_1^{10} -g(t)\,dt$. [*Hint:* Find functions like f with which you can compare g.]

2. Let $f(t)$ be the "greatest integer function"; that is, $f(t)$ is the greatest integer which is less than or equal to t—for example, $f(n) = n$ for any integer, $f(5\frac{1}{2}) = 5$, $f(-5\frac{1}{2}) = -6$, and so on.

   (a) Draw a graph of $f(t)$ on the interval $[-4, 4]$.
   (b) Find $\int_0^1 f(t)\,dt$, $\int_0^6 f(t)\,dt$, $\int_{-2}^2 f(t)\,dt$, and $\int_0^{4.5} f(t)\,dt$.
   (c) Find a general formula for $\int_0^n f(t)\,dt$, where n is any positive integer.
   (d) Let $F(x) = \int_0^x f(t)\,dt$, where $x > 0$. Draw a graph of F for $x \in [0, 4]$, and find a formula for $F'(x)$, where it is defined.

3. Suppose that $f(t)$ is piecewise constant on $[a, b]$. Let $g(t) = f(t) + k$, where k is a constant.

   (a) Show that $g(t)$ is piecewise constant.
   (b) Find $\int_a^b g(t)\,dt$ in terms of $\int_a^b f(t)\,dt$.

4. For $t \in [0, 1]$ let $f(t)$ be the first digit after the decimal point in the decimal expansion of t.

   (a) Draw a graph of f.   (b) Find $\int_0^1 f(t)\,dt$.

5. Using the definition of the integral, find $\int_0^1 (1 - x)\,dx$.

6. Suppose that $f(t)$ is piecewise constant on $[a, b]$. Let $F(x) = \int_a^x f(t)\,dt$. Prove that if f is continuous at $x_0 \in [a, b]$, then F is differentiable at x_0 and $F'(x) = f(x)$. (See Solved Exercise 2.)

7. Show that $-3 \leqslant \int_1^2 (t^3 - 4)\,dt \leqslant 4$.

8. (a) Show that, if f is piecewise constant on $[a, b]$, then there is an adapted partition for f which is "coarsest" in the sense that it is contained in any other adapted partition.

   (b) Use the result of (a) to give a new proof of Theorem 1.

9. Let F be defined on $[0, 1]$ by

$$f(t) = \begin{cases} 1 & \text{if } t = \tfrac{1}{2}, \tfrac{1}{3}, \tfrac{1}{4}, \tfrac{1}{5}, \ldots \\ 0 & \text{otherwise} \end{cases}$$

   (a) Show that f is integrable on $[0, 1]$.
   (b) What is $\int_0^t f(s)\,ds = F(t)$?
   (c) For which values of t is $F'(t) = f(t)$?

10. Show that if f_1 and f_2 are integrable on $[a, b]$, and s_1 and s_2 are any real numbers, then $s_1 f_1 + s_2 f_2$ is integrable on $[a, b]$, and

$$\int_a^b [s_1 f_1(t) + s_2 f_2(t)]\, dt = s_1 \int_a^b f_1(t)\, dt + s_2 \int_a^b f_2(t)\, dt$$

11. (a) Show that, if f is continuous on $[a, b]$, where $a < b$, and $f(t) > 0$ for all t in $[a, b]$, and then $\int_a^b f(t)\, dt > 0$.

    (b) Show that the result in (a) still holds if f is continuous, $f(t) \geqslant 0$ for all t in $[a, b]$, and $f(t) > 0$ for *some* t in $[a, b]$.

    (c) Are the results in (a) and (b) still true if the hypothesis of continuity is replaced by integrability?

12. Find functions f_1 and f_2, neither of which is integrable on $[0, 1]$ such that:

    (a) $f_1 + f_2$ is integrable on $[0, 1]$.

    (b) $f_1 - f_2$ is not integrable on $[0, 1]$.

13. Let f be defined on $[0, 1]$ by

$$f(t) = \begin{cases} 0 & t = 0 \\[2mm] \dfrac{1}{\sqrt{t}} & 0 < t \leqslant 1 \end{cases}$$

    (a) Show that there are no upper sums for f on $[0, 1]$ and hence that f is not integrable.

    (b) Show that every number less than 2 is not lower sum. [*Hint*: Use step functions which are zero on an interval $[0, \epsilon)$ and approximate f very closely on $[\epsilon, 1]$. Take ϵ small and use the integrability of f on $[\epsilon, 1]$.]

    (c) Show that no number greater than or equal to 2 is a lower sum. [*Hint*: Show $\epsilon f(\epsilon) + \int_\epsilon^1 f(t)\, dt < 2$ for all ϵ in $(0, 1)$.]

    (d) If you had to assign a value to $\int_0^1 f(t)\, dt$, what value would you assign?

14. Modeling your discussion after Problem 13, find the upper and lower sums for each of the following functions on $[0, 1]$. How would you define $\int_0^1 f(t)\, dt$ in each case?

    (a) $f(t) = \begin{cases} 0 & t = 0 \\[2mm] -\dfrac{1}{\sqrt[3]{t}} & t > 0 \end{cases}$

    (b) $f(t) = \begin{cases} 0 & t = 0 \\[2mm] \dfrac{1}{t^2} & t > 0 \end{cases}$

12   The Fundamental Theorem of Calculus

The fundamental theorem of calculus reduces the problem of integration to anti-differentiation, i.e., finding a function F such that $F' = f$. We shall concentrate here on the proof of the theorem, leaving extensive applications for your regular calculus text.

Statement of the Fundamental Theorem

Theorem 1 Fundamental Theorem of Calculus: Suppose that the function F is differentiable everywhere on $[a, b]$ and that F' is integrable on $[a, b]$. Then

$$\int_a^b F'(t)\,dt = F(b) - F(a)$$

In other words, if f is integrable on $[a, b]$ and F is an *antiderivative* for f, i.e., if $F' = f$, then

$$\int_a^b f(t)\,dt = F(b) - F(a)$$

Before proving Theorem 1, we will show how easy it makes the calculation of some integrals.

Worked Example 1 Using the fundamental theorem of calculus, compute $\int_a^b t^2\,dt$.

Solution We begin by finding an antiderivative $F(t)$ for $f(t) = t^2$; from the power rule, we may take $F(t) = \frac{1}{3}t^3$. Now, by the fundamental theorem, we have

$$\int_a^b t^2\, dt = \int_a^b f(t)\, dt = F(b) - F(a) = \tfrac{1}{3}b^3 - \tfrac{1}{3}a^3$$

We conclude that $\int_a^b t^2\, dt = \tfrac{1}{3}(b^3 - a^3)$ It is possible to evaluate this integral "by hand," using partitions of $[a, b]$ and calculating upper and lower sums, but the present method is much more efficient.

According to the fundamental theorem, it does not matter which anti-derivative we use. But in fact, we do not need the fundamental theorem to tell us that if F_1 and F_2 are both antiderivatives of f on $[a, b]$, then

$$F_1(b) - F_1(a) = F_2(b) - F_2(a)$$

To prove this we use the fact that any two antiderivatives of a function differ by a constant. (See Corollary 3 of the mean value theorem, Chapter 7.) We have, therefore, $F_1(t) = F_2(t) + C$, where C is a constant, and so

$$F_1(b) - F_1(a) = [F_2(b) + C] - [F_2(a) + C]$$

The C's cancel, and the expression on the right is just $F_2(b) - F_2(a)$.

Expressions of the form $F(b) - F(a)$ occur so often that it is useful to have a special notation for them: $F(t)\big|_a^b$ means $F(b) - F(a)$. One also writes $F(t) = \int f(t)\, dt$ for the antiderivative (also called an indefinite integral). In terms of this new notation, we can write the formula of the fundamental theorem of calculus in the form:

$$\int_a^b f(t)\, dt = F(t)\,\bigg|_a^b \quad \text{or} \quad \int_a^b f(t)\, dt = \left(\int f(t)\, dt\right)\bigg|_a^b$$

where F is an antiderivative of f on $[a, b]$.

Worked Example 2 Find $(t^3 + 5)\big|_2^3$.

Solution Here $F(t) = t^3 + 5$ and

$$(t^3 + 5)\big|_2^3 = F(3) - F(2)$$
$$= 3^3 + 5 - (2^3 + 5)$$
$$= 32 - 13 = 19$$

Worked Example 3 Find $\int_2^6 (t^2 + 1)\, dt$.

Solution By the sum and power rules for antiderivatives, an antiderivative for $t^2 + 1$ is $\tfrac{1}{3}t^3 + t$. By the fundamental theorem

$$\int_{2}^{6} (t^2 + 1)\, dt = \left(\frac{1}{3}t^3 + t\right)\Bigg|_{2}^{6}$$

$$= \frac{6^3}{3} + 6 - \left(\frac{2^3}{3} + 2\right)$$

$$= 78 - 4\tfrac{2}{3} = 73\tfrac{1}{3}$$

Solved Exercises

1. Evaluate $\int_0^1 x^4\, dx$.

2. Find $\int_0^3 (t^2 + 3t)\, dt$.

3. Suppose that $v = f(t)$ is the velocity at time t of an object moving along a line. Using the fundamental theorem of calculus, interpret the integral $\int_a^b v\, dt = \int_a^b f(t)\, dt$.

Exercises

1. Find $\int_a^b s^4\, ds$.

2. Find $\int_{-1}^1 (t^4 + t^{917})\, dt$.

3. Find $\displaystyle\int_0^{10} \left(\frac{t^4}{100} - t^2\right)\, dt$

Proof of the Fundamental Theorem

We will now give a complete proof of the fundamental theorem of calculus. The basic idea is as follows: Letting F be an antiderivative for f on $[a, b]$, we will show that if L_f and U_f are any lower and upper sums for f on $[a, b]$, then $L_f \leqslant F(b) - F(a) \leqslant U_f$. Since f is assumed to be integrable on $[a, b]$, the only number which can separate the lower sums from the upper sums in this way is the integral $\int_a^b f(t)\, dt$. It will follow that $F(b) - F(a)$ must equal $\int_a^b f(t)\, dt$.

To show that every lower sum is less than or equal to $F(b) - F(a)$, we must take any piecewise constant g on $[a, b]$ such that $g(t) \leqslant f(t)$ for all t in (a, b) and show that $\int_a^b g(t)\, dt \leqslant F(b) - F(a)$. Let $(t_0, t_1, \ldots, t_n)$ be a partition adapted to g and let k_i be the value of g on (t_{i-1}, t_i). Since $F' = f$, we have

$$k_i = g(t) \leqslant f(t) = F'(t)$$

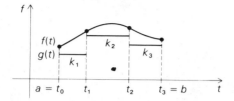

Fig. 12-1 A lower sum for the integral of f.

Hence

$$k_i \leqslant F'(t)$$

for all t in (t_{i-1}, t_i). (See Fig. 12-1). It follows from Corollary 1 of the mean value theorem (see p. 174) that

$$k_i \leqslant \frac{F(t_i) - F(t_{i-1})}{t_i - t_{i-1}}$$

Hence

$$k_i \, \Delta t_i \leqslant F(t_i) - F(t_{i-1})$$

Summing from $i = 1$ to n, we get

$$\sum_{i=1}^{n} k_i \, \Delta t_i \leqslant \sum_{i=1}^{n} [F(t_i) - F(t_{i-1})]$$

The left-hand side is just $\int_a^b g(t)\,dt$, by the definition of the integral of a step function. The right-hand side is a telescoping sum equal to $F(t_n) - F(t_0)$. (See Fig. 12-2.) Thus we have

$$\int_a^b g(t)\,dt \leqslant F(b) - F(a)$$

which is what we wanted to prove.

In the same way (see Exercise 4), we can show that if $h(t)$ is a piecewise constant function such that $f(t) \leqslant h(t)$ for all t in (a, b), then

$$F(b) - F(a) \leqslant \int_a^b h(t)\,dt$$

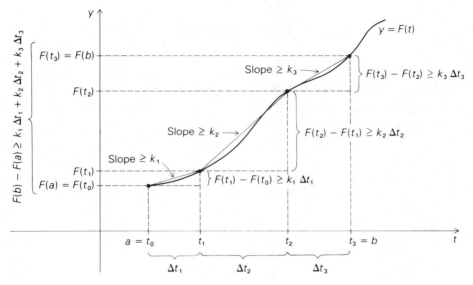

Fig. 12-2 The total change of F on $[a, b]$ is the sum of changes on the subintervals $[t_{i-1}, t_i]$. The change on the ith subinterval is at least $k_i \Delta t_i$.

as required. This completes the proof of the fundamental theorem.

Solved Exercises

4. Suppose that F is continuous on $[0, 2]$, that $F'(x) < 2$ for $0 \leqslant x \leqslant \frac{1}{3}$, and that $F'(x) < 1$ for $\frac{1}{3} \leqslant x \leqslant 2$. What can you say about $F(2) - F(0)$?

Exercises

4. Prove that if $h(t)$ is a piecewise constant function on $[a, b]$ such that $f(t) \leqslant h(t)$ for all $t \in (a, b)$, then $F(b) - F(a) \leqslant \int_a^b h(t) \, dt$, where F is any antiderivative for f on $[a, b]$.

5. Let $a_0, \ldots, a_n$ be any numbers and let $\delta_i = a_i - a_{i-1}$. Let $b_k = \Sigma_{i=1}^h \delta_i$ and let $d_i = b_i - b_{i-1}$. Express the b's in terms of the a's and the d's in terms of the δ's.

Alternative Version of the Fundamental Theorem

We have seen that the fundamental theorem of calculus enables us to compute integrals by using antiderivatives. The inverse relationship between integration and differentiation is completed by the following alternative version of the fundamental theorem, which enables us to build up an antiderivative for a function by taking definite integrals and letting the endpoint vary.

Theorem 2 Fundamental Theorem of Calculus: Alternative Version.
Let f be continuous on the interval I and let a be a number in I. Define the function F on I by

$$F(t) = \int_a^t f(s)\,ds$$

Then $F'(t) = f(t)$; that is

$$\frac{d}{dt}\int_a^t f(s)\,ds = f(t)$$

In particular, every continuous function has an antiderivative.

Proof We use the method of rapidly vanishing functions from Chapter 3. We need to show that the function

$$r(t) = F(t) - F(t_0) - f(t_0)(t - t_0)$$

is rapidly vanishing at t_0. Substituting the definition of F and using additivity of the integral, we obtain

$$r(t) = \int_a^t f(s)\,ds - \int_a^{t_0} f(s)\,ds - f(t_0)(t - t_0)$$

$$= \int_{t_0}^t f(s)\,ds - f(t_0)(t - t_0)$$

Given any number $\epsilon > 0$, there is an interval I about t_0 such that $f(t_0) - (\epsilon/2) < f(s) < f(t_0) + (\epsilon/2)$ for all s in I. (Here we use the continuity of f at t_0.) For $t > t_0$ in I, we thus have

$$\left(f(t_0) - \frac{\epsilon}{2}\right)(t - t_0) \leqslant \int_{t_0}^{t} f(s)\,ds \leqslant \left(f(t_0) + \frac{\epsilon}{2}\right)(t - t_0)$$

Subtracting $f(t_0)(t - t_0)$ everywhere gives

$$-\frac{\epsilon}{2}(t - t_0) \leqslant r(t) \leqslant \frac{\epsilon}{2}(t - t_0)$$

or

$$\left|r(t)\right| \leqslant \frac{\epsilon}{2}\left|t - t_0\right| \quad \text{(since for } t > t_0,\ |t - t_0| = t - t_0)$$

For $t < t_0$ in I, we have

$$\left(f(t_0) - \frac{\epsilon}{2}\right)(t_0 - t) \leqslant \int_{t}^{t_0} f(s)\,ds \leqslant \left(f(t_0) + \frac{\epsilon}{2}\right)(t_0 - t)$$

We may rewrite the integral as $-\int_{t_0}^{t} f(s)\,ds$. Subtracting $f(t_0)(t_0 - t)$ everywhere in the inequalities above gives

$$-\frac{\epsilon}{2}(t_0 - t) \leqslant -r(t) \leqslant \frac{\epsilon}{2}(t_0 - t)$$

so, once again,

$$\left|r(t)\right| \leqslant \frac{\epsilon}{2}\left|t - t_0\right| \quad \text{(since for } t \leqslant t_0,\ |t - t_0| = t_0 - t)$$

We have shown that, for $t \neq t_0$ in I, $|r(t)| \leqslant (\epsilon/2)|t - t_0|$; since $(\epsilon/2)\,|t - t_0| < \epsilon|t - t_0|$, Theorem 2 of Chapter 3 tells us that $r(t)$ is rapidly vanishing at t_0.

Since the proof used the continuity of f only at t_0, we have the following more general result.

Corollary Let f be integrable on the interval $I = [a, b]$ and let t_0 be a number in (a, b). If f is continuous at the point t_0, then the function $F(t) = \int_a^t f(s)\,ds$ is differentiable at t_0, and $F'(t_0) = f(t_0)$.

This corollary is consistent with the results of Problem 9, Chapter 11.

Solved Exercises

5. Suppose that f is continuous on the real line and that g is a differentiable function. Let $F(x) = \int_0^{g(x)} f(t)\,dt$. Calculate $F'(x)$.

6. Let $F(x) = \int_1^{x^2} (1/t)\,dt$. What is $F'(x)$?

Exercises

6. (a) Without using logarithms, show that $\int_1^{x^2} (1/t)\,dt = 2 \int_1^x (1/t)\,dt$.

   (b) What is the relation between $\int_1^{x^n}(1/t)\,dt$ and $\int_1^x (1/t)\,dt$?

7. Prove a fundamental theorem for $G(t) = \int_t^b f(s)\,ds$.

8. Find $\dfrac{d}{dx} \displaystyle\int_{g(x)}^{h(x)} f(t)\,dt$.

Problems for Chapter 12 ▬▬▬▬▬▬▬▬▬▬▬▬▬▬▬

1. Evaluate the following definite integrals:
   (a) $\int_1^3 t^3\,dt$ (b) $\int_{-1}^2 (t^4 + 8t)\,dt$
   (c) $\int_0^{-4} (1 + x^2 - x^3)\,dx$ (d) $\int_1^2 4\pi r^2\,dr$
   (e) $\int_2^{-1} (1 + t^2)^2\,dt$

2. If f is integrable on $[a, b]$, show by example that $F(t) = \int_a^t f(t)\,dt$ is continuous but need not be differentiable.

3. Evaluate:
   (a) $(d/dt) \int_0^t 3/(x^4 + x^3 + 1)^6\,dx$ (b) $(d/dt) \int_t^3 x^2(1 + x)^5\,dx$

   (c) $\dfrac{d}{dt} \displaystyle\int_t^4 \dfrac{u^4}{(u^2 + 1)^3}\,du$

4. Let f be continuous on the interval I and let a_1 and a_2 be in I. Define the functions:

$$F_1(t) = \int_{a_1}^t f(s)\,ds \quad \text{and} \quad F_2(t) = \int_{a_2}^t f(s)\,ds$$

   (a) Show that F_1 and F_2 differ by a constant.
   (b) Express the constant $F_2 - F_1$ as an integral.

5. Suppose that

$$f(t) = \begin{cases} t^2 & 0 \leqslant t \leqslant 1 \\ 1 & 1 \leqslant t < 5 \\ (t+6)^2 & 5 \leqslant t \leqslant 6 \end{cases}$$

   (a) Draw a graph of f on the interval $[0,6]$.
   (b) Find $\int_0^6 f(t)\,dt$.
   (c) Find $\int_0^6 f(x)\,dx$.
   (d) Let $F(t) = \int_0^t f(s)\,ds$. Find a formula for $F(t)$ in $[0,6]$ and draw a graph of F.
   (e) Find $F'(t)$ for t in $(0,6)$.

6. Prove Theorem 2 without using rapidly vanishing functions, by showing directly that $f(t_0)$ is a transition point between the slopes of lines overtaking and overtaken by the graph of F at t_0.

7. (a) Find $(d/dx)\int_1^{ax}(1/t)\,dt$, where a is a positive constant.
   (b) Show that $\int_1^{ax}(1/t)\,dt - \int_1^x(1/t)\,dt$ is a constant.
   (c) What is the constant in (b)?
   (d) Show without using logarithms that if $F(x) = \int_1^x(1/t)\,dt$, then $F(xy) = F(x) + F(y)$.
   (e) Show that if $F_c(x) = \int_1^x(c/t)\,dt$, where c is a constant, then $F_c(xy) = F_c(x) + F_c(y)$, both with and without using logarithms.

13 Limits and the Foundations of Calculus

We have developed some of the basic theorems in calculus without reference to limits. However limits are very important in mathematics and cannot be ignored. They are crucial for topics such as infinite series, improper integrals, and multi-variable calculus. In this last section we shall prove that our approach to calculus is equivalent to the usual approach via limits. (The going will be easier if you review the basic properties of limits from your standard calculus text, but we shall neither prove nor use the limit theorems.)

Limits and Continuity

Let f be a function defined on some open interval containing x_0, except possibly at x_0 itself, and let l be a real number. There are two definitions of the statement

$$\lim_{x \to x_0} f(x) = l$$

Condition 1

1. Given any number $c_1 < l$, there is an interval (a_1, b_1) containing x_0 such that $c_1 < f(x)$ if $a_1 < x < b_1$ and $x \neq x_0$.

2. Given any number $c_2 > l$, there is an interval (a_2, b_2) containing x_0 such that $c_2 > f(x)$ if $a_2 < x < b_2$ and $x \neq x_0$.

Condition 2 Given any positive number ϵ, there is a positive number δ such that $|f(x) - l| < \epsilon$ whenever $|x - x_0| < \delta$ and $x \neq x_0$.

Depending upon circumstances, one or the other of these conditions may be easier to use. The following theorem shows that they are interchangeable, so either one can be used as the definition of $\lim_{x \to x_0} f(x) = l$.

Theorem 1 *For any given f, x_0, and l, condition 1 holds if and only if condition 2 does.*

Proof (a) *Condition 1 implies condition 2.* Suppose that condition 1 holds, and let $\epsilon > 0$ be given. To find an appropriate δ, we apply condition 1, with $c_1 = l - \epsilon$ and $c_2 = l + \epsilon$. By condition 1, there are intervals (a_1, b_1) and (a_2, b_2) containing x_0 such that $l - \epsilon < f(x)$ whenever $a_1 < x < b_1$ and $x \neq x_0$, and $l + \epsilon > f(x)$ whenever $a_2 < x < b_2$ and $x \neq x_0$. Now let δ be the smallest of the positive numbers $b_1 - x_0, x_0 - a_1, b_2 - x_0$, and $x_0 - a_2$. (See Fig. 13-1.) Whenever $|x - x_0| < \delta$ and $x \neq x_0$, we have

$$a_1 < x < b_1 \quad \text{and} \quad x \neq x_0 \quad \text{so} \quad l - \epsilon < f(x) \tag{1}$$

and

$$a_2 < x < b_2 \quad \text{and} \quad x \neq x_0 \quad \text{so} \quad l + \epsilon > f(x) \tag{2}$$

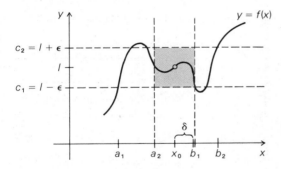

Fig. 13-1 When $|x - x_0| < \delta$ and $x \neq x_0$, $|f(x) - l| - \epsilon$.

Statements **(1)** and **(2)** together say that $l - \epsilon < f(x) < l + \epsilon$, or $|f(x) - l| < \epsilon$, which is what was required.

(b) *Condition 2 implies condition 1.* Suppose that condition 2 holds, and let $c_1 < l$ and $c_2 > l$ be given. Let ϵ be the smaller of the two positive numbers $l - c_1$ and $c_2 - l$. By condition 2, there is a positive number δ such that $|f(x) - l| < \epsilon$ whenever $|x - x_0| < \delta$ and $x \neq x_0$. Now we can verify parts 1 and 2 of condition 1, with $a_1 = b_1 = x_0 - \delta$ and $a_2 = b_2 = x_0 + \delta$. If $x_0 - \delta < x < x_0 + \delta$ and $x \neq x_0$, then $|x - x_0| < \delta$ and $x \neq x_0$, so we have $|f(x) - l| < \epsilon$; that is, $l - \epsilon < f(x) < l + \epsilon$. But this implies that $c_1 < f(x)$ and $f(x) < c_2$ (see Fig. 13-2).

The following theorem shows that our definition of continuity can be phrased in terms of limits.

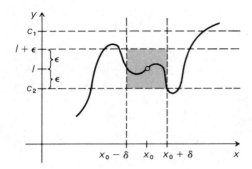

Fig. 13-2 When
$x \in (x_0 - \delta, x_0 + \delta)$ and
$x \neq x_0, c_1 < f(x) < c_2$.

Theorem 2 *Let f be defined on an open interval containing x_0. Then f is continuous at x_0 if and only if*

$$\lim_{x \to x_0} f(x) = f(x_0)$$

Proof The definition of continuity given in Chapter 3 is exactly condition 1 for the statement $\lim_{x \to x_0} f(x) = f(x_0)$

Corollary *The function f is continuous at x_0 if and only if, for every positive number ϵ, there is a positive number δ such that $|f(x) - f(x_0)| < \epsilon$ whenever $|x - x_0| < \delta$.*

Proof We have simply replaced the statement $\lim_{x \to x_0} f(x) = f(x_0)$ by its condition 2 definition. (We do not need to require that $x \neq x_0$; if $x = x_0$, $|f(x) - f(x_0)| = 0$, which is certainly less than ϵ.)

Solved Exercises

1. Using the fact that the function $f(x) = \tan x$ is continuous, show that there is a number $\delta > 0$ such that $|\tan x - 1| < 0.001$ whenever $|x - \pi/4| < \delta$.

2. Let f be defined on an open interval containing x_0, except perhaps at x_0 itself. Let

$$A = \{c \mid \text{there is an open interval } I \text{ about } x_0 \text{ such that } x \in I, x \neq x_0 \text{ implies}$$
$$f(x) > c\}$$

$B = \{d \mid$ there is an open interval J about x_0 such that $x \in J, x \neq x_0$ implies $f(x) < d\}$.

Prove that $\lim_{x \to x_0} f(x) = l$ if and only if l is a transition point from A to B.

Exercises

1. Show that there is a positive number δ such that

$$\left| \frac{x-4}{x+4} - \frac{1}{3} \right| < 10^{-6}$$

whenever $|x - 8| < \delta$.

2. Is Theorem 2 valid for f with a domain which does not contain an interval about x_0? What is the definition of limit in this case?

3. Prove that limits are unique by using the definition, Solved Exercise 2, and a theorem about transitions.

4. Which of the following functions are continuous at 0?

   (a) $f(x) = x \sin\frac{1}{x}, \quad x \neq 0, \quad f(0) = 0$

   (b) $f(x) = \frac{1}{x}\sin x, \quad x \neq 0, \quad f(0) = 0$

   (c) $f(x) = \frac{x^2}{\sin x}, \quad x \neq 0, \quad f(0) = 0$

The Derivative as a Limit of Difference Quotients

We recall the definition of the derivative given in Chapter 1.

Definition Let f be a function whose domain contains an open interval about x_0. We say that the number m_0 is the *derivative of f at x_0* provided that:

1. For every $m < m_0$, the function

$$f(x) - [f(x_0) + m(x - x_0)]$$

changes sign from negative to positive at x_0.

2. For every $m > m_0$, the function

$$f(x) - [f(x_0) + m(x - x_0)]$$

changes sign from positive to negative at x_0.

If such a number m_0 exists, we say that f is *differentiable* at x_0 and we write $m_0 = f'(x_0)$.

We will now prove that our definition of the derivative coincides with the definition found in most calculus books.

Theorem 3 *Let f be a function whose domain contains an open interval about x_0. Then f is differentiable at x_0 with derivative m_0 if and only if*

$$\lim_{x \to x_0} [f(x_0 + \Delta x) - f(x_0)] / \Delta x$$

exists and equals m_0.

Proof We will use the condition 1 form of the definition of limit. Suppose that $\lim_{x \to x_0} [f(x_0 + \Delta x) - f(x_0)] / \Delta x = m_0$. To verify that $f'(x_0) = m_0$, we must study the sign change at x_0 of $r(x) = f(x) - [f(x_0) + m(x - x_0)]$ and see how it depends on m.

First assume that $m < m_0$. Since the limit of difference quotients is m_0, there is an interval (a, b) containing zero such that

$$m < \frac{f(x_0 + \Delta x) - f(x_0)}{\Delta x}$$

whenever $a < \Delta x < b$, $\Delta x \neq 0$. Writing x for $x_0 + \Delta x$, we have

$$m < \frac{f(x) - f(x_0)}{x - x_0} \tag{3}$$

whenever $x_0 + a < x < x_0 + b$, $x \neq x_0$—that is, whenever $x_0 + a < x < x_0$ or $x_0 < x < x_0 + b$.

In case $x_0 + a < x < x_0$, we have $x - x_0 < 0$, and so equation (3) can be transformed to

$$m(x - x_0) > f(x) - f(x_0)$$
$$0 > f(x) - [f(x_0) + m(x - x_0)]$$

When $x_0 < x < x + b$, we have $x - x_0 > 0$, so equation (3) becomes

$$m(x - x_0) < f(x) - f(x_0)$$
$$0 < f(x) - [f(x_0) + m(x - x_0)]$$

In other words, $f(x) - [f(x_0) + m(x - x_0)]$ changes sign from negative to positive at x_0.

Similarly, if $m > m_0$, we can use part 2 of the condition 1 definition of limit to show that $f(x) - [f(x_0) + m(x - x_0)]$ changes sign from positive to negative at x_0. This completes the proof that $f'(x_0) = m_0$.

Next we show that if $f'(x_0) = m_0$, then $\lim_{\Delta x \to 0} [f(x_0 + \Delta x) - f(x_0)]/\Delta x = m_0$. This is mostly a matter of reversing the steps in the first half of the proof, with slightly different notation. Let $c_1 < m_0$. To find an interval (a, b) containing zero such that

$$c_1 < \frac{f(x_0 + \Delta x) - f(x_0)}{\Delta x} \tag{4}$$

whenever $a < \Delta x < b$, $\Delta x \neq 0$, we use the fact that $f(x) - [f(x_0) + c_1(x - x_0)]$ changes sign from negative to positive at x_0. There is an interval (a_1, b_1) containing x_0 such that $f(x) - [f(x_0) + c_1(x - x_0)]$ is negative when $a_1 < x < x_0$ and positive when $x_0 < x < b_1$. Let $a = a_1 - x_0 < 0$ and $b = b_1 - x_0 > 0$. If $a < \Delta x < 0$, we have $a_1 < x_0 + \Delta x < x_0$, and so

$$0 > f(x_0 + \Delta x) - [f(x_0) + c_1 \Delta x]$$
$$c_1 \Delta x > f(x_0 + \Delta x) - f(x_0)$$

$$c_1 < \frac{f(x_0 + \Delta x) - f(x_0)}{\Delta x} \quad (\text{since } \Delta x < 0)$$

which is just equation (4). If $0 < \Delta x < b$, we have $x_0 < x_0 + \Delta x < b_1$, and so

$$0 < f(x_0 + \Delta x) - [f(x_0) + c_1 \Delta x]$$
$$c_1 \Delta x < f(x_0 + \Delta x) - f(x_0)$$

$$c_1 < \frac{f(x_0 + \Delta x) - f(x_0)}{\Delta x}$$

which is equation (4) again.

Similarly, if $c_2 > m_0$, there is an interval (a, b) containing zero such that $c_2 > [f(x_0 + \Delta x) - f(x_0)]/\Delta x$ whenever $a < \Delta x < b$, $\Delta x \neq 0$. Thus we have shown that $\lim_{\Delta x \to 0} [f(x_0 + \Delta x) - f(x_0)]/\Delta x = m_0$.

Combining Theorems 1 and 3, we can now give an ϵ-δ characterization of the derivative.

Corollary *Let f be defined on an open interval containing x_0. Then f is differentiable at x_0 with derivative $f'(x_0)$ if and only if, for every positive number ϵ, there is a positive number δ such that*

$$\left| \frac{f(x_0 + \Delta x) - f(x_0)}{\Delta x} - f'(x_0) \right| < \epsilon$$

whenever $|\Delta x| < \delta$ and $\Delta x \neq 0$.

Proof We have just rephrased the statement .

$$\lim_{\Delta x \to 0} \frac{f(x_0 + \Delta x) - f(x_0)}{\Delta x} = f'(x_0)$$

using the ϵ-δ definition of limit.

Solved Exercises

---

3. If f is differentiable at x_0, what is $\lim_{x \to x_0} [f(x) - f(x_0)]/(x - x_0)$?

4. Let f be defined near x_0, and define the function $g(\Delta x)$ by

$$g(\Delta x) = \begin{cases} \dfrac{f(x_0 + \Delta x) - f(x_0)}{\Delta x} & \Delta x \neq 0 \\ \\ m_0 & \Delta x = 0 \end{cases}$$

where m_0 is some number.
   Show that $f'(x_0) = m_0$ if and only if g is continuous at 0.

Exercises

---

5. Find $\lim_{x \to 2} (x^2 + 4x + 3 - 15)/(x - 2)$.

6. Prove Theorem 3 using the ϵ-δ definition of the derivative, and draw pictures to illustrate your constructions.

7. (a) Suppose that $f'(x_0) = g'(x_0) \neq 0$. Find $\lim\limits_{x \to x_0} \dfrac{f(x) - f(x_0)}{g(x) - g(x_0)}$.

   (b) Find $\lim\limits_{x \to 1} \dfrac{2x^3 - 2}{3x^2 - 3}$.   (c) Find $\lim\limits_{x \to 1} \dfrac{x^n - 1}{x^m - 1}$.

8. Evaluate $\lim\limits_{x \to 0} \dfrac{\sqrt{1 - x^2} - 1}{x}$:

   (a) By recognizing the limit to be a derivative.

   (b) By rationalizing.

9. Evaluate the following limit by recognizing the limit to be a derivative:

$$\lim_{x \to \pi/4} \frac{\sin x - (\sqrt{2}/2)}{x - (\pi/4)}$$

The Integral as a Limit of Sums

In this section, we shall need the notion of a limit of a sequence. (See your calculus text for examples and discussion.)

Definition Let $a_1, a_2, a_3, \ldots$ be a sequence of real numbers and let l be a real number. We say that l is the *limit* of the sequence and we write $\lim\limits_{n \to \infty} a_n = l$ if, for every $\epsilon > 0$, there is a number N such that $|a_n - l| < \epsilon$ for all $n \geqslant N$.

Now let f be defined on an interval $[a, b]$. In our definition of the integral $\int_a^b f(t)\,dt$ in Chapter 12, we considered partitions $(t_0, t_1, \ldots, t_n)$ of $[a, b]$ and lower sums:

$$\sum_{i=1}^{n} k_i(t_i - t_{i-1}) \quad \text{where } k_i < f(t) \text{ for all } t \text{ in } (t_{i-1}, t_i)$$

and upper sums:

$$\sum_{i=1}^{n} l_i(t_i - t_{i-1}) \quad \text{where } f(t) < l_i \text{ for all } t \text{ in } (t_{i-1}, t_i)$$

The integral was then defined to be the transition point between upper and lower sums, i.e., that number S, if it exists, for which every $s < S$ is a lower sum and every $s > S$ is an upper sum.

We also may consider sums of the form

$$S_n = \sum_{i=1}^{n} f(c_i)(t_i - t_{i-1}) \quad \text{where } c_i \text{ in } [t_{i-1}, t_i]$$

called *Riemann sums*. The integral can be defined as a limit of Riemann sums. This is reasonable since any Riemann sum associated to a given partition lies between any upper and lower sums for that partition. (See Fig. 13-3.) The following is a precise statement, showing that the limit approach coincides with the method of exhaustion.

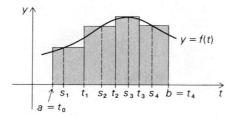

Fig. 13-3 Illustrating a Riemann sum.

Theorem 4   *Let f be a bounded function on $[a, b]$.*

1. *Assume that f is integrable and that the maximum of the numbers $\Delta t_i = t_i - t_{i-1}$ goes to zero as $n \to \infty$. Then for any choice of c_i,*

$$\lim_{n \to \infty} S_n = \int_a^b f(t)\, dt$$

2. *Suppose that for every choice of c_i and t_i with the maximum of Δt_i tending to zero as $n \to \infty$, the limit $\lim_{n \to \infty} S_n = S$. Then f is integrable with integral S.*

Let U and L denote the set of upper and lower sums for f, respectively. We have shown in Chapter 12 that if l is in L and u is in U, then $l \leqslant u$ and that $L = (-\infty, S_1)$ or $(-\infty, S_1]$ and $U = (S_2, \infty)$ or $[S_2, \infty)$. Integrability amounts to the requirement that $S_1 = S_2$. We shall need the following lemma before we proceed to the proof of Theorem 4.

Lemma Let $a = s_0 < s_1 < \cdots < s_N = b$ be a partition of $[a, b]$ and let $\epsilon > 0$ be given. If $a = t_0 < t_1 < \cdots < t_n = b$ is any partition with $\Delta t_i < \epsilon/N$ for every i, then the total length of the intervals $[t_{i-1}, t_i]$ which are not contained entirely within some (s_{p-1}, s_p) is less than ϵ.

Proof Since there are just N points in the s-partition, there are no more than $2N$ intervals $[t_{i-1}, t_i]$ which contain points of the s-partition (it is $2N$ because s_i could be in two such intervals by being a common endpoint). Since each such interval has length strictly less than $\epsilon/2N$, the total length is $(2N)(\epsilon/2N) = \epsilon$.

Proof of Theorem 4 We give the proof of part 1. Part 2 is left for the reader (see Problem 13). Let ϵ be given and let $|f(x)| \leqslant M$ for all t in $[a, b]$. Choose piecewise constant functions $g(t) \leqslant f(t)$ and $h(t) \geqslant f(t)$ such that

$$\int_a^b h(t)\,dt - \int_a^b g(t)\,dt < \frac{\epsilon}{2}$$

(so both integrals are within $\epsilon/2$ of $\int_a^b f(t)\,dt$) which is possible since f is integrable.

Let $u_0 < u_1 < \cdots < u_r$ and $v_0 < v_1 < \cdots < v_s$ be adapted partitions for g and h. Let $s_0 < s_1 < \cdots < s_N$ be a partition adapted for both g and h obtained by taking all the u's and v's together. Let $g(x) = k_p$ on (s_{p-1}, s_p) and $h(x) = l_p$ on (s_{p-1}, s_p).

Choose N_1 so that $\Delta t_i < \dfrac{(\epsilon/2M)}{2N}$, if $n > N_1$; this is possible since the maximum of the Δt_i goes to zero as $n \to \infty$.

By the lemma, the total length of the intervals $[t_{i-1}, t_i]$ not contained in some (s_{p-1}, s_p) is less than $\epsilon/2M$. Thus

$$S_n = \sum_{i=1}^n f(c_i)\Delta t_i$$

$$= \sum_{\substack{i \text{ such that} \\ [t_{i-1}, t_i] \\ \text{lies in some} \\ (s_{p-1}, s_p)}} f(c_i) \, \Delta t_i + \sum_{\substack{\text{rest of} \\ \text{the } i\text{'s}}} f(c_i) \, \Delta t_i$$

$$\leqslant \sum_{p=1}^{N} l_p \, \Delta s_p + M \cdot \frac{\epsilon}{2M}$$

$$= \int_a^b h(t) \, dt + \frac{\epsilon}{2}$$

$$\leqslant \int_a^b f(t) \, dt + \frac{\epsilon}{2} + \frac{\epsilon}{2}$$

$$= \int_a^b f(t) \, dt + \epsilon$$

In a similar way we show that $S_n \geqslant \int_a^b f(t) \, dt - \epsilon$. Thus, if $n \geqslant N_1$, then $|S_n - \int_a^b f(t) \, dt| < \epsilon$, so our result is proved.

Solved Exercises

5. Let $S_n = \sum_{i=1}^{n} (1 + i/n)$. Prove that $S_n \to \frac{3}{2}$ as $n \to \infty$ (a) directly and (b) using Riemann sums.

6. Use Theorem 4 to demonstrate that $\int_a^b [f(x) + g(x)] \, dx = \int_a^b f(x) \, dx + \int_a^b g(x) \, dx$. (You may assume that the limit of a sum is the sum of the limits.)

Exercises

10. (a) Prove that

$$\lim_{\pi \to \infty} \left(\frac{1}{n+1} + \frac{1}{n+2} + \cdots + \frac{1}{n+n} \right) = \int_0^1 \frac{dx}{1+x} = \ln 2$$

(b) Evaluate the sum for $n = 10$ and compare with $\ln 2$.

11. Use Theorem 4 to demonstrate the following:

(a) $\int_a^b cf(x) \, dx = c \int_a^b f(x) \, dx$

(b) If $f(x) \leqslant g(x)$ for all x in $[a, b]$, then $\int_a^b f(x)\, dx \leqslant \int_a^b g(x)\, dx$.

12. Let f be a bounded function on $[a, b]$. Assume that for every $\epsilon > 0$ there is a $\delta > 0$ such that, if $a = t_0 < t_1 < \cdots < t_n = b$ is a partition with $\Delta t_i < \delta$, and c_i is any point in $[t_{i-1}, t_i]$, then

$$\left| \sum_{i=1}^n f(c_i) \Delta t_i - S \right| < \epsilon$$

Prove that f is integrable with integral S.

Problems for Chapter 13

1. Evaluate $\lim\limits_{x \to \pi^2} (\cos \sqrt{x} + 1)/(x - \pi^2)$ by recognizing the limit to be a derivative.

2. Evaluate:

$$\lim_{x \to \pi^2} \left[\frac{\sin \sqrt{x}}{(\sqrt{x} - \pi)(\sqrt{x} + \pi)} + \tan \sqrt{x} \right]$$

3. Using Theorem 2 and the limit laws, prove that if f and g are continuous at x_0, then so are $f + g$, fg, and f/g (if $g(x_0) \neq 0$).

4. Prove the chain rule, $(f \cdot g)'(x_0) = f'(g(x_0)) \cdot g'(x_0)$, via limits as follows:

   (a) Let $y = g(x)$ and $z = f(y)$, and write

   $$\Delta y = g'(x_0)\Delta x + \rho(x)$$

   Show that

   $$\lim_{\Delta x \to 0} \frac{\rho(x)}{\Delta x} = 0$$

   Also write

   $$\Delta z = f'(y_0)\Delta y + \sigma(y) \qquad y_0 = g(x_0)$$

   and show that

   $$\lim_{\Delta y \to 0} \frac{\sigma(y)}{\Delta y} = 0$$

   (b) Show that

   $$\Delta z = f'(y_0)g'(x_0)\Delta x + f'(y_0)\rho(x) + \sigma(g(x))$$

   (c) Note that $\sigma(g(x)) = 0$ if $\Delta y = 0$. Thus show that

$$\frac{\sigma(g(x))}{\Delta x} = \begin{cases} \dfrac{\sigma(g(x))}{\Delta y}\dfrac{\Delta y}{\Delta x} & \text{if } \Delta y \neq 0 \\[2ex] 0 & \text{if } \Delta y = 0 \end{cases} \rightarrow 0$$

as $\Delta x \rightarrow 0$.

(d) Use parts (b) and (c) to show that $\lim_{\Delta x \to 0} \Delta z / \Delta x = f'(y_0)g'(x_0)$.

5. Write down Riemann sums for the given functions. Sketch.

(a) $f(x) = x/(x + 1)$ for $1 \leqslant x \leqslant 6$ with $t_0 = 1, t_1 = 2, t_2 = 3, t_3 = 4, t_4 = 5, t_5 = 6; c_i = i - 1$ on $[i - 1, i]$.

(b) $f(x) = x + \sin[(\pi/2)x]$, $0 \leqslant x \leqslant 6$ with $t_i = i, i = 0, 1, 2, 3, 4, 5, 6$. Find Riemann sums S_6 with

$c_i = i$ on $[i - 1, i]$

and

$c_i = i - \frac{1}{2}$ on $[i - 1, i]$

6. Write each of the following integrals as a limit.

(a) $\int_1^3 [1/(x^2 + 1)]\, dx$; partition $[1, 3]$ into n equal parts and use a suitable choice of c_i.

(b) $\int_0^\pi (\cos\frac{1}{2}x + x)\, dx$; partition $[0, \pi]$ into n equal parts and use a suitable choice of c_i.

7. Let

$$S_n = \sum_{i=1}^{n} \left(\frac{i}{n} + \frac{i^2}{n^2}\right)\frac{1}{n}$$

Prove that $S_n \rightarrow \frac{5}{6}$ as $n \rightarrow \infty$ by using Riemann sums.

8. Expressing the following sums as Riemann sums, show that:

(a) $\displaystyle\lim_{n\to\infty} \sum_{i=1}^{n} \left[\sqrt{\frac{i}{n}} - \left(\frac{i}{n}\right)^{3/2}\right]\frac{1}{n} = \frac{4}{15}$

(b) $\displaystyle\lim_{n\to\infty} \sum_{i=1}^{n} \frac{3n}{(2n + i)^2} = \frac{1}{2}$

9. Write down a Riemann sum for $f(x) = x^3 + 2$ on $[-2, 3]$ with $t_i = -2 + (i/2); i = 0, 1, 2, \ldots, 10$.

10. Write $\int_{-\pi/4}^{\pi/4} (1 + \tan x)\, dx$ as a limit. (Partition $[-\pi/4, \pi/4]$ into $2n$ equal parts and choose c_i appropriately.)

11. Use Theorem 4 to prove that $\int_a^c f(x)\, dx = \int_a^b f(x)\, dx + \int_b^c f(x)\, dx$.

12. Show that

$$f''(x_0) = \lim_{\Delta x \to 0} \frac{f(x_0 + 2\,\Delta x) + f(x_0) - 2f(x_0 + \Delta x)}{(\Delta x)^2}$$

if f'' is continuous at x_0.

13. Prove part 2 of Theorem 4 using the following outline (demonstrate each of the statements). Let f be a bounded function on the interval $[p, q]$, and let ϵ be any positive number. Prove that there are real numbers m and M and numbers x_m and x_M in $[p, q]$ such that:

    1. $m \leqslant f(x) \leqslant M$ for all x in $[p, q]$

    2. $f(x_m) < m + \epsilon$ and $f(x_M) > M - \epsilon$

[*Hint*: Let S be the set of real numbers z such that $f(x) \leqslant z \leqslant f(y)$ for some x and y in $[p, q]$. Prove that S is an interval by using the completeness axiom.]

14. Prove that $e = \lim\limits_{h \to 0} (1 + h)^{1/h}$ using the following outline. Write down the equation $\ln'(1) = 1$ as a limit and substitute into $e = e^1$. Use the continuity of e^x and $e^{\ln y} = y$.

15. (a) A function f defined on a domain D is called *uniformly continuous* if for any $\epsilon > 0$ there is a $\delta > 0$ such that $|x - y| < \delta$ implies $|f(x) - f(y)| < \epsilon$. Show that a continuous function on $[a, b]$ is uniformly continuous. (You may wish to use the proof of Theorem 3, Chapter 11 for inspiration.)

    (b) Use (a) to show that a continuous function on $[a, b]$ is integrable.

16. (Cauchy Sequences.) A sequence $a_1, a_2, a_3, \ldots$ is called a *Cauchy sequence* if for every $\epsilon > 0$ there is a number N such that $|a_n - a_m| < \epsilon$ whenever $n \geqslant N$ and $m \geqslant N$. Prove that every convergent sequence is a Cauchy sequence.

17. Use the following outline to prove that every Cauchy sequence $a_1, a_2, a_3, \ldots$ converges to some real number.

    (a) Using the definition of a Cauchy sequence, with $\epsilon = 1$, prove that the sequence is bounded.

    (b) Let S be the set of real numbers x such that $a_n < x$ for infinitely many n. Prove that S is an interval of the form (l, ∞) or $[l, \infty)$. (Use the completeness axiom.)

    (c) Prove that $\lim\limits_{n \to \infty} a_n = l$.

Appendix: Solutions

CHAPTER 1

1. No. The polynomial $f(x) = x^2 - 2x + 1 = (x-1)^2$ has a root at 1, but it does not change sign there, since $(x-1)^2 > 0$ for all $x \neq 1$.

2. For n even, $x^n > 0$ for all $x \neq 0$, so there is no sign change. For n odd, x^n is negative for $x < 0$ and positive for $x > 0$, so there is a sign change from negative to positive at zero.

3. As in Worked Example 1, the quadratic $(x - r_1)(x - r_2)$ changes sign from positive to negative at the smaller root and from negative to positive at the larger root. Thus the sign change at r_1 is from negative to positive if $r_1 > r_2$ and from positive to negative if $r_1 < r_2$.

4. First we must find a motion $y = 2x + b$ which passes through $y = \frac{1}{2}$ when $x = 1$. We find $\frac{1}{2} = 2 \cdot 1 + b$, or $b = -\frac{3}{2}$. Now we look at the difference
$$\tfrac{1}{2}x^2 - (2x - \tfrac{3}{2}) = \tfrac{1}{2}x^2 - 2x + \tfrac{3}{2} = \tfrac{1}{2}(x^2 - 4x + 3) = \tfrac{1}{2}(x - 3)(x - 1)$$
The factor $\frac{1}{2}(x - 3)$ is negative near $x = 1$, so $\frac{1}{2}(x - 3)(x - 1)$ changes sign from positive to negative at 1. It follows that the "test" object with uniform velocity 2 passes our moving object, so its velocity is at most 2.

5. We must study the sign changes at $x_0 = 0$ of $x^3 - mx = x(x^2 - m)$. If $m < 0$, the factor $x^2 - m$ is everywhere positive and the product $x(x^2 - m)$ changes sign from negative to positive at $x_0 = 0$. If $m > 0$, then $x^2 - m$ is negative for x in $(-\sqrt{m}, \sqrt{m})$, so the sign change of $x(x^2 - m)$ at $x_0 = 0$ is from positive to negative. The number $m_0 = 0$ fits the definition of the derivative, so the derivative at $x_0 = 0$ of $f(x) = x^3$ is zero. The tangent line at $(0,0)$ has slope zero, so it is just the x axis. (See Fig. S-1-1.)

6. The equation of the tangent line at $(x_0, f(x_0))$ is
$$y = f(x_0) + f'(x_0)(x - x_0)$$
If $x_0 = 3$, $f(3) = 2$, and $f'(3) = \sqrt[5]{8}$, we get
$$y = 2 + \sqrt[5]{8}(x - 3) = \sqrt[5]{8}x + (2 - 3\sqrt[5]{8})$$
The y intercept is $2 - 3\sqrt[5]{8}$.

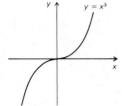

Fig. S-1-1 The tangent line at $(0,0)$ to $y = x^3$ is the x axis.

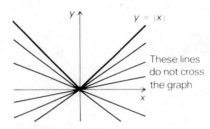

These lines do not cross the graph

Fig. S-1-2 The graph $y = |x|$ has no tangent line at $(0,0)$.

7. The graph of $f(x) = |x|$ is shown in Fig. S-1-2. None of the lines through $(0,0)$ with slopes between -1 and 1 cross the graph at $(0,0)$, so there can be no m_0 satisfying the definition of the derivative.

8. By the second definition on p. 8, the velocity is the derivative of x^2 at $x = 3$. This derivative was calculated in Worked Example 4, it is 6.

9. $f'(x) = 3$ for all x, so $f'(8) = 3$.

10. The velocity at time t is $f'(t)$, where $f(t) = 4.9t^2$. We have $f'(t) = 2(4.9)t = 9.8t$; at $t = 3$, this is 29.4 meters per second. The acceleration is $f''(t) = 9.8$ meters per second per second.

11. $f'(x) = 2 \cdot 3x + 4 = 6x + 4$, so $f'(1) = 10$. Also, $f(1) = 9$, so the equation of the tangent line is $y = 9 + 10(x - 1)$, or $y = 10x - 1$.

12. Let $f(x) = ax^2 + bx + c$. Then $f'(x) = 2ax + b$ and the derivative of this is $f''(x) = 2a$. Hence $f''(x)$ is equal to zero when $a = 0$—that is, when $f(x)$ is a linear function $bx + c$.

CHAPTER 2

1. The set B ($x^2 - 1 > 0$) consists of the intervals $(1, \infty)$ and $(-\infty, -1)$. From Fig. S-2-1, we see that -1 is a transition point from B to A and that 1 is a transition point from A to B.

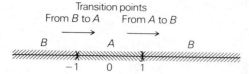

Transition points
From B to A From A to B

B A B

-1 0 1

Fig. S-2-1 Transitions between intervals occur at common endpoints.

2. $-1/1000$ is the transition point from A to C, and $1/1000$ is the transition point from C to B. There are no other transition points; in particular, the transition from A to B has a "gap." (You should draw yourself a figure like the one above for this situation.)

3. The state line crossing.

4. $1/x$ is negative for $x < 0$ and positive for $x > 0$; thus, $1/x$ changes from negative to positive at 0. (Note that $1/x$ is not defined for $x = 0$.)

5. (a) We begin by drawing a picture and locating on it the intersection point of the two lines (Fig. S-2-2). To satisfy the definition of overtaking, we must find an interval I such that conditions 1 and 2 on p. 21 hold. From the picture, we guess that $I = (-\infty, \infty)$ will work. For condition 1, we must show that, if $x < 0$, then $3x + 2 < 2x + 2$. This is a simple chain of inequalities:

$x < 0$ implies

$3x < 2x$ implies

$3x + 2 < 2x + 2.$

Similarly, for condition 2, we show that $x > 0$ implies $3x + 2 > 2x + 2$.

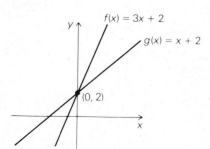

$f(x) = 3x + 2$

$g(x) = x + 2$

$(0, 2)$

Fig. S-2-2 f overtakes g at 0.

(b) To use Theorem 3, we merely note that both lines pass through $(0, 2)$ and that the slope 3 of f_1 is larger than the slope 2 of f_2. Theorem 2 then assures us that f_1 overtakes f_2 at 0. (You may notice that the calculation in (a) looks very much like the proof of Theorem 3. This is no accident. The proofs of theorems are often modeled after particular calculations, with letters substituted for specific values.)

6. We make the sketch first (Fig. S-2-3).

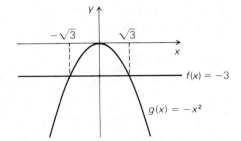

Fig. S-2-3 *f* overtakes *g* at $\sqrt{3}$.

It looks as if f overtakes g at $\sqrt{3}$, and that an interval which works should be $I = (-\sqrt{3}, \infty)$. Let us verify conditions 1 and 2. If $x \in I$ and $x < \sqrt{3}$, we have $-\sqrt{3} < x < \sqrt{3}$. Is $-3 < -x^2$ for these values of x? In other words, is $x^2 - 3 < 0$? In other words, do $(x - \sqrt{3})$ and $(x + \sqrt{3})$ have opposite signs? Yes, they do, because $x - \sqrt{3} < 0$ and $x + \sqrt{3} > 0$. That verifies condition 1. For condition 2 we notice that, if $x > \sqrt{3}$, then $x > -\sqrt{3}$ as well, so $x^2 - 3 = (x - \sqrt{3})(x + \sqrt{3})$ is positive; i.e., $-x^2 < -3$.

7. From a sketch (Fig. S-2-4), we suspect that g overtakes f at the first point where the two graphs intersect, i.e., at the smaller root of the equation $2x^2 = 5x - 3$. Since $2x^2 - 5x + 3 = (2x - 3)(x - 1)$, we guess that g overtakes f at 1. An interval which works should be $(-\infty, \frac{3}{2})$. We must verify conditions 1 and 2.

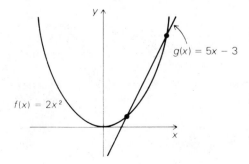

Fig. S-2-4 *g* overtakes *f* at 1.

To check condition 1, we must show that if $x < 1$ and $x \in I$, then $5x - 3 < 2x^2$. But this is the same as $2x^2 - 5x + 3 > 0$; i.e., $(2x - 3)(x - 1) > 0$; i.e., $2x - 3$ and $x - 1$ have the same sign. But they are both negative if $x < 1$, so condition 1 is satisfied. To check condition 2, we must show that, if $x > 1$ and $x < \frac{3}{2}$, then $5x - 3 > 2x^2$; i.e., $(2x - 3)(x - 1)$ have opposite signs. They do, since $2x - 3 = 2(x - \frac{3}{2}) < 0$ while $x - 1 > 0$.

8. We have $f(x_0) = 2$. To construct A and B, we must look at all the lines through $(x_0, f(x_0)) = (0, 2)$ and determine which ones overtake and are overtaken by the graph of f at 0. The equation of the general (nonvertical) line through $(0, 2)$ is $y = mx + 2$. In Fig. S-2-5, we plot the parabola $y = f(x) = x^2 + \frac{3}{2}x + 2$ and the lines $y = mx + 2$ for $m = 0, 1$, and 2.

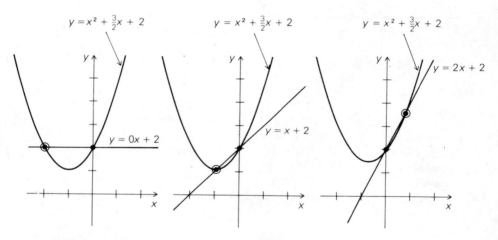

Fig. S-2-5 The line $y = x + 2$ is overtaken by $y = x^2 + \frac{3}{2}x + 2$ at $x = 0$ and $y = 2x + 2$ overtakes $y = x^2 + \frac{3}{2}x + 2$.

We see from the figure that the lines $y = 0x + 2$ and $y = x + 2$ are overtaken at $x_0 = 0$ by the parabola $y = x^2 + \frac{3}{2}x + 2$, while the line $y = 2x + 2$ overtakes the parabola at 0. Thus, the derivative $f'(0)$, if it exists, must lie somewhere between 1 and 2. To find the precise value of $f'(0)$, we make another observation from the figure: a line is overtaken by the parabola when the second (circled) point of intersection lies to the left of 0, while the line overtakes the parabola when the second point of intersection is to the right of 0. We can locate this point for arbitrary m by solving the equation

$$x^2 + \frac{3}{2}x + 2 = mx + 2$$
$$x^2 + (\frac{3}{2} - m)x = 0$$

$$x(x + \tfrac{3}{2} - m) = 0$$
$$x = 0 \quad \text{or} \quad x = m - \tfrac{3}{2}$$

The circled point is the one which is *not* 0; i.e., it is $m - \tfrac{3}{2}$. For $m < \tfrac{3}{2}$, $m - \tfrac{3}{2}$ is negative, and the parabola overtakes the line; for $m > \tfrac{3}{2}$, $m - \tfrac{3}{2}$ is positive, and the line overtakes the parabola. Thus $A = (-\infty, \tfrac{3}{2})$ and $B = (\tfrac{3}{2}, \infty)$. The point of transition between A and B is $\tfrac{3}{2}$; we conclude that $f(x) = x^2 + \tfrac{3}{2}x + 2$ is differentiable at $x_0 = 0$ and that $f'(0) = \tfrac{3}{2}$.

CHAPTER 3

1.

x	$g_1(x) = x$	$g_2(x) = x^2$	$g_3(x) = x^3$
0.1	0.1	0.01	0.001
0.01	0.01	0.0001	0.000001
0.002	0.002	0.000004	0.000000008
0.0004	0.0004	0.00000016	0.000000000064

The functions g_2 and g_3 appear to vanish rapidly at 0. (In fact, g_2 does by the quadratic function rule; see Exercise 3 for g_3.)

2. (a) If f and g vanish at x_0, then $f(x_0) = g(x_0) = 0$. Then $(f + g)(x_0) = f(x_0) + g(x_0) = 0 + 0 = 0$.

   (b) $fg(x_0) = f(x_0)g(x_0) = 0 \cdot g(x_0) = 0$.

3. Let $f(x) = (x - x_0)^2$. $f(x_0) = (x_0 - x_0)^2 = 0^2 = 0$. To show that $f'(x_0) = 0$, we multiply out: $f(x) = x^2 - 2x_0 x + x_0^2$. We may apply the quadratic function rule to get $f'(x) = 2x - 2x_0$; therefore, $f'(x_0) = 2x_0 - 2x_0 = 0$.

4. We use both parts of Theorem 3. By part 2, $(-1)r_2$ is rapidly vanishing at x_0. By part 1, $r_1 + (-1)r_2 = r_1 - r_2$ is rapidly vanishing at x_0.

5. By Theorem 2, $r(x) = f(x) - f(x_0) - f'(x_0)(x - x_0)$ vanishes rapidly at x_0. By Theorem 3, so does

   $$ar(x) = af(x) - af(x_0) - af'(x_0)(x - x_0)$$

   Thus, by Theorem 2, $af(x)$ is differentiable at x_0 with derivative $af'(x_0)$.

6. Referring to Fig. 3-3, we see that the extreme values of y for points in the bow-tie region occur at the four points which are marked in the figure. If a and b are the endpoints of I, the y values at these points are

   $$(f'(x_0) - 1)(a - x_0)$$
   $$(f'(x_0) + 1)(a - x_0)$$

$$(f'(x_0) - 1)(b - x_0)$$
$$(f'(x_0) + 1)(b - x_0)$$

Choose B so that all four of these numbers lie between $-B$ and B.

7. Since we now have the product rule, we write $f(x) = x^4 = x^2 \cdot x^2$ and compute $f'(x) = 2x^2 \cdot x^2 + x^2 \cdot 2x^2 = 4x^3$. Then $f'(0) = f(0) = 0$, so $f(x)$ vanishes rapidly at 0.

8. Let $f(x) = (x - 3)^2 (x - 7)^2$. At $x = 3$, $(x - 3)^2$ vanishes rapidly (see Solved Exercise 3). By Theorem 5, so does $(x - 3)^2 (x - 7)^2$. Similarly for $x = 7$. (Alternatively, you may compute $f'(x)$ by the product rule.)

9. Say that the line l_1 has slope $g'(x_0) + 1$. This line is then $y = g(x_0) + (g'(x_0) + 1)(x - x_0)$. It intersects $y = \frac{1}{2}g(x_0)$ when

$$\tfrac{1}{2}g(x_0) = g(x_0) + (g'(x_0) + 1)(x - x_0)$$

i.e., when

$$x = x_0 - \frac{1}{2}\frac{g(x_0)}{g'(x_0) + 1}$$

If $g'(x_0) + 1 > 0$, this point is to the left of x_0. The left-hand endpoint I is this value of x, or the left-hand endpoint of the interval which works for l_1 overtaking the graph of g, whichever is nearer to x_0. If $g'(x_0) + 1$ is negative, we may simply use the endpoint of the interval which works. (Draw your own figure for this case.)

10. Let

$$g(x) = \begin{cases} 1 & \text{if } x \geqslant 0 \\ 0 & \text{if } x < 0 \end{cases}$$

The beginning of the proof of Theorem 7 shows that such a g cannot be differentiable at 0.

CHAPTER 4

1. We have

$$0 + 0 = 0 \qquad \text{by I.1}$$

so

$$x(0 + 0) = x \cdot 0$$
$$x \cdot 0 + x \cdot 0 = x \cdot 0 \qquad \text{by II.5}$$
$$[x \cdot 0 + x \cdot 0] + [-(x \cdot 0)] = x \cdot 0 + [-(x \cdot 0)]$$

$x \cdot 0 + \{x \cdot 0 + [-(x \cdot 0)]\} = x \cdot 0 + [-(x \cdot 0)]$ by I.2

$x \cdot 0 + 0 = 0$ by I.4

$x \cdot 0 = 0$ by I.3

2. $(x + y)^2 = (x + y)(x + y)$
$$= (x + y)x + (x + y)y \qquad \text{by II.5}$$
$$= (x^2 + yx) + (xy + y^2) \qquad \text{by II.5}$$
$$= x^2 + yx + xy + y^2 \qquad \text{by I.2}$$
$$= x^2 + xy + xy + y^2 \qquad \text{by II.1}$$
$$= x^2 + 1 \cdot xy + 1 \cdot xy + y^2 \quad \text{by II.1 and II.3}$$
$$= x^2 + (1 + 1)xy + y^2 \qquad \text{by I.5}$$
$$= x^2 + 2xy + y^2$$

Note that the associativity properties of addition and multiplication allow us to write unparenthesized sums and products of more than two numbers, like $x^2 + yx + xy + y^2$ or $2xy$, without ambiguity.

As your experience in writing proofs increases, you may begin to omit explicit references to the reasons for your steps, or to do more than one manipulation at each step, but for the moment you should write out all the details.

3. We assume, of course, that $6 = 5 + 1 = 4 + 1 + 1 = 3 + 1 + 1 + 1$, etc. Now
$$2 \cdot 3 = (1 + 1) \cdot (1 + 1 + 1)$$
$$= 1 \cdot (1 + 1 + 1) + 1 \cdot (1 + 1 + 1) \qquad \text{II.5}$$
$$= 1 \cdot 1 + 1 \cdot 1 + 1 \cdot 1 + 1 \cdot 1 + 1 \cdot 1 + 1 \cdot 1 \quad \begin{array}{l}\text{II.5, extended to sums with} \\ \text{several terms}\end{array}$$
$$= 1 + 1 + 1 + 1 + 1 + 1 \qquad \text{II.3}$$
$$= 2 + 1 + 1 + 1 + 1$$
$$= 3 + 1 + 1 + 1 = 4 + 1 + 1 = 5 + 1 = 6$$

4. $(-x) \cdot y + xy = [(-x) + x] \cdot y \qquad \text{II.5}$
$$= 0 \cdot y \qquad \text{I.1 and I.4}$$
$$= 0 \qquad \text{Solved Exercise 1}$$

So

$(-x) \cdot y + xy + [-(xy)] = 0 + [-(xy)]$

$(-x) \cdot y + 0 = 0 + [-(xy)] \qquad \text{I.4}$

$(-x) \cdot y = -(xy) \qquad \text{I.3}$

5. $\left(\dfrac{a}{b} + \dfrac{c}{d}\right)(bd) = \left(a \cdot \dfrac{1}{b} + c \cdot \dfrac{1}{d}\right) \cdot (bd)$

$$= a \cdot \frac{1}{b} \cdot bd + c \cdot \frac{1}{d} \cdot bd \qquad \text{II.5}$$

$$= a \cdot 1 \cdot d + c \cdot 1 \cdot b \qquad \text{II.1 and II.4}$$

$$= ad + bc \qquad \text{II.1 and II.3}$$

Now $bd \neq 0$, since if $bd = 0$, then as $b \neq 0$, we would have $d = (1/b) \cdot bd = (1/b) \cdot 0 = 0$, contradicting the assumption that $d \neq 0$. Thus, we may multiply both sides of the equation above by $1/bd$ to get

$$\frac{a}{b} + \frac{c}{d} = (ad + bc)\left(\frac{1}{bd}\right) = \frac{ad + bc}{bd}$$

6. By III.4, the only possibility other than $0 < 1$ is $1 \leqslant 0$. We will show that this leads to a contradiction. If

$$1 \leqslant 0$$

then

$$1 + (-1) \leqslant 0 + (-1) \qquad \text{III.5}$$
$$0 \leqslant -1$$
$$0 \leqslant (-1) \cdot (-1) \qquad \text{III.6}$$
$$0 \leqslant 1$$

Thus, $0 \leqslant 1$ and $1 \leqslant 0$. By III.3, this implies that $0 = 1$, contradicting II.3.

7. Since $c \leqslant 0$, $c + (-c) \leqslant -c$, so $0 \leqslant -c$. Also, $x \leqslant y$, so $x + (-x) \leqslant y + (-x)$, i.e., $0 \leqslant (y - x)$. By III.6, we have

$$0 \leqslant (-c)(y - x)$$
$$0 \leqslant -cy + cx$$
$$cy + 0 \leqslant cy + (-cy) + cx \qquad \text{III.5}$$
$$cy \leqslant cx$$
$$cx \geqslant cy$$

8. By III.4, the only possibility other than $a \leqslant b$ is $b < a$. By the assumed property of a and b, this implies $b < b$, which is impossible. Thus, $b < a$ is impossible, so we must have $a \leqslant b$.

9. By III.4, we have that either $x \geqslant 0$ or $x \leqslant 0$. If $x \geqslant 0$, we have $x^2 = x \cdot x \geqslant 0$ by III.6. If $x \leqslant 0$, then $(-1) \cdot x \geqslant (-1) \cdot 0$, by Solved Exercise 7, so $-x \geqslant 0$. By III.6, we have $0 \leqslant (-x)(-x) = (-1)^2 \cdot x^2 = x^2$. So, in either case, we have $x^2 \geqslant 0$.

10. Suppose that x_1 and x_2 are in $[a, b)$, i.e., $a \leqslant x_1 < b$ and $a \leqslant x_2 < b$, and that $x_1 < y < x_2$. We must prove that $a \leqslant y < b$. But $a \leqslant x_1$ and $x_1 < y$ imply $a \leqslant y$, and $y < x_2$ and $x_2 < b$ imply $y < b$. Thus $y \in [a, b)$.

11. To show that $c^2 = 2$, we will show that neither $c^2 < 2$ nor $c^2 > 2$ is possible. Suppose $c^2 < 2$. If $0 < h < 1$, we have $0 < h^2 < h \cdot 1 = h$, and $(c + h)^2 = c^2 + 2ch + h^2 < c^2 + 2ch + h = c^2 + (2c + 1)h$. If $0 < h < (2 - c^2)/(2c + 1)$ (there always exists such an h, since $2 - c^2 > 0$), we have

$$(c + h)^2 < c^2 + (2c + 1) \cdot \frac{2 - c^2}{2c + 1} = 2$$

Thus, $c + h$ is an element of S which is larger than c, contradicting the fact that c is the upper endpoint of S.

If $c^2 > 2$, we look at $(c - h)^2$, for small positive h. This equals $c^2 - 2ch + h^2 > c^2 - 2ch$. If we choose h such that $0 < h < (c^2 - 2)/2c$, we have

$$(c - h)^2 > c^2 - 2ch > c^2 - 2c\left(\frac{c^2 - 2}{2c}\right) = c^2 - c^2 + 2 = 2$$

Thus, $c - h$ does not belong to S. This again contradicts the fact that c is the upper endpoint of S.

12. If there is a rational number whose square is 2, it can be written as m/n, where m and n are positive integers with no common factor. The equation $(m/n)^2 = 2$ implies $m^2 = 2n^2$; thus m^2 being divisible by 2, must be even. Since the square of an odd number is odd $((2k + 1)^2 = 2(2k^2 + 2k) + 1)$, m cannot be odd, so it must be even. If we write $m = 2l$, we have the equation $(2l)^2 = 2n^2$, or $4l^2 = 2n^2$, or $2l^2 = n^2$. Thus n^2, and hence n, must be even. Since m and n are both even, they have the common factor of 2, contradicting our assumption. (This argument is found in Euclid.)

13. First, we show that there are rational and irrational numbers in $(0, b - a)$. To find a rational number, let n be an integer such that $1/(b - a) < n$. (See Problem 5, p. 52.) Then $0 < 1/n < (b - a)$. For an irrational, let m be an integer such that $\sqrt{2}/(b - a) < m$. Then $0 < \sqrt{2}/m < b - a$, and $\sqrt{2}/m$ is irrational. (If $\sqrt{2}/m$ were rational, then so would be the product $m(\sqrt{2}/m) = \sqrt{2}$. But $\sqrt{2}$ is irrational, as was shown in Solved Exercise 12.

Now look at the sequence of rational numbers

$$\ldots, -\frac{3}{n}, -\frac{2}{n}, -\frac{1}{n}, 0, \frac{1}{n}, \frac{2}{n}, \frac{3}{n}, \frac{4}{n}, \ldots$$

which extends to infinity in both directions. Some of these numbers are to the left of (a, b) and others are to the right. Since the distance between successive numbers is less than the length $b - a$ of (a, b), the sequence cannot jump over (a, b), and there must be a rational of the form p/n in (a, b).

Next, look at the sequence

$$\ldots, -\frac{2\sqrt{2}}{m}, -\frac{\sqrt{2}}{m}, 0, \frac{\sqrt{2}}{m}, \frac{2\sqrt{2}}{m}, \frac{3\sqrt{2}}{m}, \ldots$$

These numbers are all irrational, since if $q\sqrt{2}/m$ were rational, so would be $\sqrt{2}$. By the same argument as for the previous sequence, one of the numbers $q\sqrt{2}/m$ must lie in (a, b).

CHAPTER 5

1. First sketch the graph of g (Fig. S-5-1). Take $\frac{1}{2}$ for c_2. The inequality $g(0) <$ $\frac{1}{2}$ is satisfied by g at $x_0 = 0$, since $g(0) = 0$, but no matter what open interval I we take about 0, there are positive numbers x in I for which $g(x) = 1$, which is greater than $\frac{1}{2}$. Since it is not possible to choose I such that condition 2 in the definition of continuity is satisfied, with $x_0 = 0$, $c = \frac{1}{2}$, it follows that g is not continuous at 0.

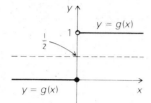

Fig. S-5-1 This step function is discontinuous at 0.

2. The graph of f is shown in Fig. S-5-2. We must establish conditions 1 and 2 in the definition of continuity. First, we check condition 2. Let c_2 be such that $f(x_0) = f(0) = 0 < c_2$, i.e., $c_2 > 0$. We must find an open interval J about 0 such that $f(x) < c_2$ for all $x \in I$. From Fig. S-5-2, we see that we should try $J = (-c_2, c_2)$. For $x \geqslant 0$ and $x \in I$, we have $f(x) = x < c_2$. For $x < 0$ and $x \in J$, we have $f(x) = -x$. Since $x > -c_2$, we have $-x < c_2$, i.e., $f(x) < c_2$. Thus, for all $x \in J$, $f(x) < c_2$. For condition 1, we have $c_1 < f(0) = 0$. We can take I to be any open interval about 0, even $(-\infty, \infty)$, since $c_1 < 0 \leqslant f(x)$ for all real numbers x. Hence f is continuous at 0.

Notice that the interval J has to be chosen smaller and smaller as $c_2 > 0$ is nearer and nearer to zero. (It is accidental to this example that the interval I can be chosen independently of c_1.)

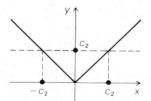

Fig. S-5-2 The absolute value function is continuous at 0.

3. If $f(x_0)$ is not zero, it is either positive or negative. Suppose first that $f(x_0) > 0$. In the definition of continuity, we may set $c_1 = 0$ in condition 1. We conclude that there is an open interval I about x_0 on which $0 < f(x)$, so $1/f(x)$ is defined on I. If $f(x_0) < 0$, we use condition 2 of the definition instead to conclude that $f(x) < 0$, and hence $1/f(x)$ is defined, for all x in some open interval J about x_0.

4. (a) The function is discontinuous at $\ldots -2, -1, 0, 1, 2, \ldots$ (Take $c_2 = \frac{1}{2}$ in condition 2 of the definition of continuity.)

   (b) The function is continuous, even though it is not differentiable at the "corners" of the graph.

   (c) The function is continuous, even though its graph cannot be drawn without removing pencil from paper. In fact, if you take any point (x_0, y_0) *on the curve,* the part of the curve lying over some open interval about x_0 *can* be drawn without removing pencil from paper.

   (d) The function is not continuous at 1. (Use condition 1 of the definition with $c_1 = \frac{3}{2}$.)

5. This is a quotient of polynomials, and the denominator takes the value $4 \neq 0$ at $x = 1$, so the function is continuous there, by part 2 of the corollary to Theorem 1.

6. The absolute value function (see Solved Exercise 2) was shown to be continuous at $x_0 = 0$. However, in Solved Exercise 7 of Chapter 1, we showed that the same function fails to be differentiable at 0. Thus, a function which is continuous at a point need not be differentiable there, so the converse of Theorem 1 is false. Notice that our example does not contradict Theorem 1; since the absolute value function is not differentiable at 0, Theorem 1 simply has nothing to say about it.

7. $f(x) = x^3 + 8x^2 + x$ is continuous at $x = 0$, since it is a polynomial and hence is differentiable. Let $c = 1/1000 > f(0) = 0$. By the definition of continuity, there is an open interval I about 0 such that $x \in I$ implies $f(x) < c$. Let δ be the right endpoint of I. Thus $0 \leqslant x < \delta$ implies $x \in I$, so $f(x) < 1/1000$.

8. Let $g(x) = f(x) + A$. Let $g(x_0) < c$, so $f(x_0) < c - A$. There is an interval I about x_0 with $f(x) < c - A$ if $x \in I$ by the definition of continuity for f at x_0. Hence if $x \in I$, $g(x) = f(x) + A < c$, so condition 1 holds for g. Condition 2 is proved similarly.

9. It is not so easy to show that T is convex. In fact, if we do not assume that $f(x) \neq d$ for all x in $[a, b]$, then T might not be convex, although S always will be. (See Fig. S-5-3 on the next page.)

10. We can let

$$f(x) = \begin{cases} \sin\dfrac{1}{x} & \text{for } x > 0 \\ 0 & \text{for } x \leqslant 0 \end{cases}$$

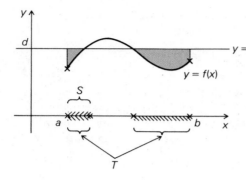

Fig. S-5-3 The set of x for which $f(x) < d$ might not be convex.

11. If $f(x) = x^5 + x^4 - 3x^2 + 2x + 8$, we have $f(100) = 10,099,970,208 > 0$, while $f(-100) = -9,900,030,192 < 0$. Since f is a polynomial, it is differentiable by the results of Chapter 3 and hence is continuous by Theorem 1. By the intermediate value theorem, there is some number c in $(-100, 100)$ such that $f(c) = 0$.

12. Suppose that $y_1 < y < y_2$, and y_1 and y_2 belong to T. To prove that $y \in T$, we begin by observing that $y_1 = f(x_1)$ and $y_2 = f(x_2)$ for some x_1 and x_2 in $[a, b]$. $f(x_1) \neq f(x_2)$ implies that $x_1 \neq x_2$, since $x_1 = x_2$ would imply $f(x_1) = f(x_2)$. Let J be the closed interval whose endpoints are x_1 and x_2. (We do not know whether $x_1 < x_2$ or $x_2 < x_1$, so J could be $[x_1, x_2]$ or $[x_2, x_1]$.) Since $f(x_1) < y < f(x_2)$, the intermediate value theorem applied to f on J tells us that $f(c) = y$ for some c in J. But c lies in $[a, b]$ as well, so $y \in T$.

13. We calculate $f'(x)$ by the quotient rule:

$$f'(x) = \frac{(x - 1) - (x + 1)}{(x - 1)^2} = -\frac{2}{(x - 1)^2}$$

$f'(0) = -2$, which is negative, so f is decreasing at 0 by Theorem 3. It is getting colder.

14. $f'(x) = 6x^2 - 18x + 12 = 6(x^2 - 3x + 2) = 6(x - 1)(x - 2)$. This is positive for x in $(-\infty, 1)$ and x in $(2, \infty)$, zero for $x = 1$ and 2, and negative for x in $(1, 2)$. By Theorem 3, we conclude that f is increasing at each x in $(-\infty, 1)$ and $(2, \infty)$ and that f is not increasing at x in $(1, 2)$ (because it is decreasing there). For $x = 1$ and $x = 2$, we cannot tell whether f is increasing or not without using other methods (discussed in Chapter 6).

15. The derivative at 0 of each of these functions is 0, so Theorem 3 does not give us any information: we must use the definition of increasing and decreasing functions.

    For $f(x) = x^3$, we have $f(x) = x \cdot x^2$. Since $x^2 \geqslant 0$ for all x, x^3 has the same sign as x; i.e., $f(x) < 0 = f(0)$ for $x < 0$ and $f(x) > 0 = f(0)$ for $x > 0$.

This means that f is increasing at 0; we can take the interval I to be $(-\infty, \infty)$. A similar argument shows that $-x^3 = -x \cdot x^2$ is decreasing at 0.

Finally, for $f(x) = x^2$, we have $f(x) \geqslant 0 = f(0)$ for all x. The conditions for increasing or decreasing can never be met, because we would need to have $f(x) < f(0)$ for x on one side or the other of 0. Thus, x^2 is neither increasing nor decreasing at 0.

16. Nothing is wrong. Although f is increasing at each point of its domain, it is not increasing (in fact, not defined) at 0, which lies in the interval $[-1,1]$, so Theorem 5 does not apply to this interval.

17. We apply Theorem $4'$ to f on the intervals $[x_1, b]$ and $[b, x_2]$, where $a < x_1 < b < x_2 < c$ concluding that $f(x) < f(b)$ for all x in $[x_1, b]$ and $f(b) < f(x)$ for all x in $[b, x_2]$. But this says exactly that f is increasing at b.

18. $f(x) = x^8 + x^4 + 8x^9 - x$ is continuous on $[0, 10,000]$, so by the boundedness lemma, f is bounded above on $[0, 10,000]$ by some number M. (The lemma doesn't tell us how to find such an M—this requires some calculation.)

19. We know from the proof of the extreme value theorem that T is an interval, and that the upper endpoint of T belongs to T. But we also know that f has a *minimum* value of T on $[a, b]$; this minimum value is the lower endpoint of T and belongs to T. Since T contains both its endpoints, it is a closed interval.

The answer to the last question is "no." For instance, consider $f(x) = x^2$ on $[-1, \frac{1}{2}]$. The set of values of f (see Fig. S-5-4) is $[0,1]$, but $f(\frac{1}{2}) = \frac{1}{4}$ is not an endpoint.

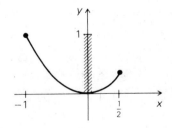

Fig. S-5-4 The value of f at an endpoint of an interval need not be an endpoint of the set of values of f on the interval.

CHAPTER 6

1. We begin by finding the critical points: $f'(x) = 12x^3 - 24x^2 + 12x = 12x(x^2 - 2x + 1) = 12x(x-1)^2$; the critical points are thus 0 and 1. Since $(x-1)^2$ is always nonnegative, the only sign change is from negative to positive at zero. Thus zero is a local minimum point, while the critical point 1 is not a turning point.

2. $(d/dx)x^n = nx^{n-1}$. This changes sign from negative to positive at zero if n is even (but $n \neq 0$); there is no sign change if n is odd. Thus zero is a local minimum point if n is even and at least 2; otherwise there is no turning point.

3. $g'(x) = -f'(x)/f(x)^2$, which is zero when $f'(x)$ is zero, negative when $f'(x)$ is positive, and positive when $f'(x)$ is negative (as long as $f(x) \neq 0$. Thus, as long as $f(x) \neq 0$, $g'(x)$ changes sign when $f'(x)$ does, but in the opposite direction. Hence we have:

   1. Local minimum points of $g(x) = $ local maximum points of $f(x)$ where $f(x) \neq 0$.

   2. Local maximum points of $g(x) = $ local minimum points of $f(x)$ where $f(x) \neq 0$.

   Note that $g(x_0)$ is not defined if $f(x_0) = 0$, so g cannot have a turning point there, even if f does.

4. $f'(x) = 3x^2 - 1$ and $f''(x) = 6x$. The critical points are zeros of $f'(x)$; that is $x = \pm(1/\sqrt{3})$; $f''(-1/\sqrt{3}) = -(6/\sqrt{3}) < 0$ and $f''(1/\sqrt{3}) = 6/\sqrt{3} > 0$. By the second derivative test, $-(1/\sqrt{3})$ is a local maximum point and $1/\sqrt{3}$ is a local minimum point.

5. Let $g = f'$. Then $g'(x_0) = 0$ and $g''(x_0) > 0$, so x_0 is a local minimum point for $g(x) = f'(x)$, and so $f'(x)$ cannot change sign at x_0. Thus x_0 cannot be a turning point for f.

6. The function $l(x) = 0$ is the linear approximation to all three graphs at zero. For $x \neq 0$, $x^4 > 0$ and $-x^4 < 0$, so x^4 is concave upward at zero and $-x^4$ is concave downward. The graph of the function x^3 crosses its tangent line at $x_0 = 0$, so x^3 can be neither concave upward nor concave downward at that point.

7. $f'(x) = 9x^2 - 8$, $f''(x) = 18x$. Thus f is concave upward when $18x > 0$ (that is, when $x > 0$) and concave downward when $x < 0$. Before sketching the graph, we notice that turning points occur when $x = \pm\sqrt{\frac{8}{9}} = \pm\frac{2}{3}\sqrt{2}$. See Fig. S-6-1.

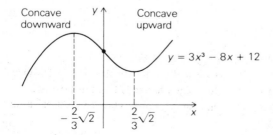

Concave downward   y   Concave upward

$y = 3x^3 - 8x + 12$

$-\frac{2}{3}\sqrt{2}$ $\frac{2}{3}\sqrt{2}$ x

Fig. S-6-1 $f''(x) = 18x$ changes sign at 0.

8. (a) If $f''(x_0) < 0$ (or $f''(x_0) > 0$) the linear approximation to f at x_0 is always greater (or less) than $f(x)$ for x near x_0.

(b) Let $f(x) = 1/x$; $f'(x) = -1/x^2$; $f''(x) = 2/x^3$. We have $f''(1) = 2$, so f is concave upward at 1. Thus, for x near 1, the linear approximation $l(x) = f(1) + f'(1)(x - 1) = 1 - 1(x - 1) = 2 - x$ is always less than $1/x$. (We leave it to the reader as an exercise to prove directly from the laws of inequalities that $2 - x \leqslant 1/x$ for all $x > 0$.)

9. For each of the functions listed, $f'(0) = f''(0) = f'''(0) = 0$. We already saw (Solved Exercise 6) that zero is a local minimum point for x^4 and a local maximum point for $-x^4$. For x^5, we have $f''(x) = 20x^3$, which changes sign from negative to positive at $x = 0$, so zero is a point of inflection at which the graph crosses the tangent line from below to above. (You can see this directly: $x^5 < 0$ for $x < 0$ and $x^5 > 0$ for $x > 0$.) Similarly, for $-x^5$ zero is an inflection point at which the graph of the function crosses the tangent line from above to below.

10. Suppose that f'' changes sign from negative to positive at x_0. If $h(x) = f(x) - [f(x_0) + f'(x_0)(x - x_0)]$, then $h'(x) = f'(x) - f'(x_0)$ and $h''(x) = f''(x)$. Since $f''(x)$ changes sign from negative to positive at x_0, so does $h''(x)$; thus x_0 is a local minimum point for $h'(x)$. Since $h'(x_0) = 0$, we have $h'(x) > 0$ for all $x \neq x_0$ in some open interval (a, b) about x_0. It follows from the corollary to Theorem 5 of Chapter 5 that h is increasing on $(a, x_0]$ and on $[x_0, b)$. Since $h(x_0) = 0$, this gives $h(x) < 0$ for $x \in (a, x_0)$ and $h(x) > 0$ for $x \in (x_0, b)$; that is, h changes sign from negative to positive at x_0. The case where f'' changes sign from positive to negative at x_0 is similar.

11. We have $f'(x) = 96x^3 - 96x^2 + 18x$, so $f''(x) = 288x^2 - 192x + 18$ and $f'''(x) = 576x - 192$. To find inflection points, we begin by solving $f''(x) = 0$; the quadratic formula gives $x = (4 \pm \sqrt{7})/12$. Using our knowledge of parabolas, we conclude that f'' changes from positive to negative at $(4 - \sqrt{7})/12$ and from negative to positive at $(4 + \sqrt{7})/12$; thus both are inflection points. (One could also evaluate $f'''((4 \pm \sqrt{7})/12)$, but this would be more complicated.)

12. $\frac{1}{2}(2x^3 + 3x^2 + x + 1) = x^3 + \frac{3}{2}x^2 + \frac{1}{2}x + \frac{1}{2}$. Substituting $x - \frac{1}{2}$ for x (since the coefficient of x^2 is $\frac{3}{2}$) gives

$(x - \frac{1}{2})^3 + \frac{3}{2}(x - \frac{1}{2})^2 + \frac{1}{2}(x - \frac{1}{2}) + \frac{1}{2}$
$= x^3 - \frac{3}{2}x^2 + \frac{3}{4}x - \frac{1}{8} + \frac{3}{2}(x^2 - x + \frac{1}{4}) + \frac{1}{2}(x - \frac{1}{2}) + \frac{1}{2}$
$= x^3 - \frac{1}{4}x + \frac{1}{2}$

Thus, after being shifted along the x and y axes, the cubic becomes $x^3 - \frac{1}{4}x$. Since $c < 0$, it is of type III.

13. For type I, $f''(x)$ is never zero. For types II_1 and II_3, there is a point
($x = 0$) at which f'' vanishes. This is not an inflection point, but rather a
point where the graph is especially flat. Notice, in Fig. 6-12, that types II_1
and II_3 represent transitions between types I and III. The flat spot on a
graph of type II is the location where a pair of inflection points appear as
we pass from type I to type III. Types II_1 and II_3 are mirror images of one
another, distinguished by the side of the local minimum point on which the
flat spot occurs.

CHAPTER 7

1. Here $a = -2$ and $b = 3$, so $[f(b) - f(a)]/(b - a) = [27 - (-8)]/$
$[3 - (-2)] = 35/5 = 7$. This should equal $f'(x_0) = 3x_0^2$, for some x_0 between -2 and 3. In fact, we must have $3x_0^2 = 7$ or $x_0 = \pm\sqrt{7/3} \approx \pm 1.527$.
Each of these values is between -2 and 3, so either will do as x_0. The
situation is sketched in Fig. S-7-1.

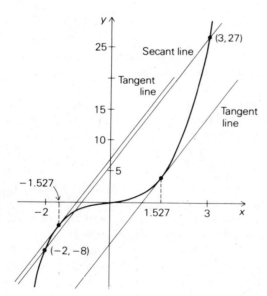

Fig. S-7-1 Tangent lines parallel to a secant line of $y = x^3$.

2. If, in Corollary 1, we take $S = (0, \infty)$, then the hypothesis $f'(x) \in S$ becomes
$f'(x) > 0$, and the conclusion $[f(x_2) - f(x_1)]/(x_2 - x_1) \in S$ becomes
$[f(x_2) - f(x_1)]/(x_2 - x_1) > 0$, or $f(x_2) > f(x_1)$ when $x_2 > x_1$. Thus
Corollary 1, in this case, is just the increasing function theorem (Theorem 5,
Chapter 5).

3. Let $f(t)$ be the position of the train at time t; let a and b be the beginning and ending times of the trip. By Corollary 1, with $S = [40,50]$, we have $40 \leqslant [f(b) - f(a)]/(b - a) \leqslant 50$. But $f(b) - f(a) = 200$, so

$$40 \leqslant \frac{200}{b - a} \leqslant 50$$

$$\frac{1}{5} \leqslant \frac{1}{b - a} \leqslant \frac{1}{4}$$

$$5 \geqslant b - a \geqslant 4$$

Hence the trip takes somewhere between 4 and 5 hours.

4. The derivative of $1/x$ is $-(1/x^2)$, so by Corollary 3, $F(x) = 1/x + C$, where C is a constant, on any interval not containing zero. So $F(x) = 1/x + C_1$ for $x < 0$ and $F(x) = 1/x + C_2$ for $x > 0$. Hence C_1 and C_2 are constants, but they are not necessarily equal.

5. Since $f(0) = 0$ and $f(2) = 0$, Rolle's theorem shows that f' is zero at some x_0 in $(0,2)$; that is, $0 < x_0 < 2$.

6. Apply the horserace theorem with $f_1(x) = f(x)$, $f_2(x) = x^2$, and $[a, b] = [0,1]$.

CHAPTER 8

1. We set $y = mx + b$ and solve for x, obtaining $x = (y/m) - (b/m)$. Thus x is completely determined by y, and the inverse function is $g(y) = (y/m) - (b/m)$. (Notice that if $m = 0$, there is no solution for x unless $y = b$; the constant function is not invertible.)

2. The domain S consists of all x such that $x \neq -d/c$. Solving $y = (ax + b)/(cx + d)$ for x in terms of y gives:

$$y(cx + d) = ax + b$$

$$(cy - a)x = -dy + b$$

If $cy - a \neq 0$, we have the unique solution

$$x = \frac{-dy + b}{cy - a} = g(y)$$

It appears that the inverse function g is defined for all $y \neq a/c$; we must, however, check the condition that $cy - a \neq 0$ when $y = (ax + b)/(cx + d)$; Note that

$$cy - a = c\left(\frac{ax + b}{cx + d}\right) - a$$

$$= \frac{cax + bc - acx - ad}{cx + d} = \frac{bc - ad}{cx + d}$$

Thus there are two cases to consider:

Case 1: $bc - ad \neq 0$. In this case, f is invertible on S, and T consists of all y which are unequal to a/c.

Case 2: $bc - ad = 0$. In this case, $b = ad/c$ and

$$y = \frac{ax + ad/c}{cx + d} = \frac{1}{c} \left(\frac{acx + ad}{cx + d} \right)$$

$$= \frac{a}{c} \left(\frac{cx + d}{cx + d} \right) = \frac{a}{c}$$

So f is a constant function, which is not invertible.

3. Writing $y = x^2 + 2x + 1 = (x + 1)^2$, we get $x = -1 \pm \sqrt{y}$. In this case, there is *not* a unique solution for x in terms of y; in fact, for $y > 0$ there are two solutions, while for $y < 0$ there are no solutions. Looking at the graph of f (Fig. S-8-1), we see that the range of values of f is $[0, \infty]$ and that f becomes invertible if we restrict it to $(-\infty, -1]$ or $[-1, \infty)$. Since we want an interval containing zero, we choose $[-1, \infty)$. Now if $x \geq -1$ in the equation $x = -1 \pm \sqrt{y}$, we must choose the positive square root, so the inverse function g has domain $T = [0, \infty)$ and is defined by $g(y) = -1 + \sqrt{y}$. (See Fig. S-8-2.) Hence $g(9) = -1 + \sqrt{9} = -1 + 3 = 2$. (Note that $f(2) = 2^2 + 2 \cdot 2 + 1 = 9$.) Also, $g(x) = -1 + \sqrt{x}$. (Again, one just substitutes. It is possible to use any letter to denote the variable in a function.)

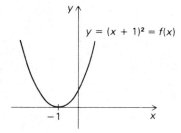

Fig. S-8-1 How can we restrict f to make it invertible?

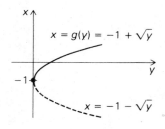

Fig. S-8-2 The inverse of the function in Figure S-8-1.

4. The graphs, which we obtain by viewing the graphs in Fig. 8-3 from the reverse side of the page, are shown in Fig. S-8-3.

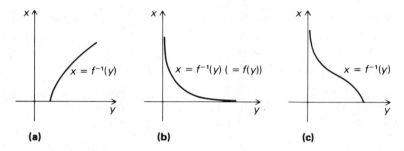

(a)   **(b)**   **(c)**

Fig. S-8-3 Inverses of the functions in Figure 8-3.

5. A simple test is the following: The function is invertible if each horizontal line meets the graph in at most one point. Applying this test, we find that the functions (a) and (c) are invertible while (b) is not.

6. (a) $f'(x) = 5x^4 + 1 > 0$, so, by Theorem 1, f is invertible on $[-2,2]$. The domain of the inverse is $[f(-2), f(2)]$, which is $[-34, 34]$.

   (b) Since $f'(x) > 0$ for all x in $(-\infty, \infty)$, f is increasing on $(-\infty, \infty)$. Now f takes arbitrarily large positive and negative values as x varies over $(-\infty, \infty)$; it takes all values in between by the intermediate value theorem, so the domain of f^{-1} is $(-\infty, \infty)$. There is no simple formula for $f^{-1}(y)$, the solution of $x^5 + x = y$, in terms of y, but we can calculate $f^{-1}(y)$ for any specific values of y to any desired degree of accuracy. (This is really no worse than the situation for $\sqrt{x}$. If the inverse function to $x^5 + x$ had as many applications as the square root function, we would learn about it in high school, tables would be readily available for it, calculators would calculate it at the touch of a key, and there would be a standard notation like $\sqrt[\mathcal{L}\!\mathcal{O}]{y}$ for the solution of $x^5 + x = y$, just as $\sqrt[5]{y}$ is the standard notation for the solution of $x^5 = y$.)*

   (c) Since $f(1) = 1^5 + 1 = 2$, $f^{-1}(2)$ must equal 1.

   (d) To calculate $f^{-1}(3)$—that is, to seek an x such that $x^5 + x = 3$—we use the *method of bisection*: Since $f(1) = 2 < 3$ and $f(2) = 34 > 3$, x must lie between 1 and 2. We can squeeze toward the correct answer by successively testing $f(x)$ at the midpoint of the interval in which x lies to obtain a new interval of half the length containing x:

---

*For further discussion of this point, see E. Kasner and J. Newman, *Mathematics and the Imagination*, Simon and Schuster, 1940, pp. 16-18.

$$f(1.5) = 9.09375 \quad \text{so } 1 < x < 1.5$$
$$f(1.25) = 4.30176 \quad \text{so } 1 < x < 1.25$$
$$f(1.1) = 2.71051 \quad \text{so } 1.1 < x < 1.25$$
$$f(1.15) = 3.16135 \quad \text{so } 1.1 < x < 1.15$$
$$f(1.14) = 3.06541 \quad \text{so } 1.1 < x < 1.14$$
$$f(1.13) = 2.97244 \quad \text{so } 1.13 < x < 1.14$$
$$f(1.135) = 3.01856 \quad \text{so } 1.13 < x < 1.135$$

Thus, to two decimal places, $x = 1.13$. (About 10 minutes of further experimentation gave $f(1.132997566) = 3.000000002$ and $f(1.132997565) = 2.999999991$. What does this tell you about $f^{-1}(3)$?)

7. We find $f'(x) = 5x^4 - 1$, which has roots at $\pm\sqrt[4]{\frac{1}{5}}$. It is easy to check that each of these roots is a turning point, so f is invertible on $(-\infty, -\sqrt[4]{\frac{1}{5}}]$, $[-\sqrt[4]{\frac{1}{5}}, \sqrt[4]{\frac{1}{5}}]$, and $[\sqrt[4]{\frac{1}{5}}, \infty)$.

8. Here $f'(x) = nx^{n-1}$. Since n is odd, $n - 1$ is even and $f'(x) > 0$ for all $x \neq 0$. Thus f is increasing on $(-\infty, \infty)$ and so it is invertible there. Since $f(x)$ is continuous and x^n takes arbitrarily large positive and negative values, it must take on all values in between, so the range of values of f, which is the domain of f^{-1}, is $(-\infty, \infty)$. We usually denote $f^{-1}(y)$ by $y^{1/n}$ or $\sqrt[n]{y}$. We have just proved that if n is odd, every real number has a unique real nth root.

9. We find that $f'(x) = nx^{n-1}$ changes sign at zero if n is even. Thus f is invertible on $(-\infty, 0]$ and on $[0, \infty)$. Since x^n is positive for all n and takes arbitrarily large values, the range of values for f is $[0, \infty)$, whether the domain is $(-\infty, 0]$ or $[0, \infty)$. The inverse function to f on $[0, \infty)$ is usually denoted by $y^{1/n} = \sqrt[n]{y}$; the inverse function to f on $(-\infty, 0]$ is denoted by $-y^{1/n}$ or $-\sqrt[n]{y}$. Remember that these are defined only for $y \geq 0$.

10. We find $y = f(x) = (ax + b)/(cx + d)$, so

$$f'(x) = \frac{a(cx + d) - c(ax + b)}{(cx + d)^2} = \frac{ad - bc}{(cx + d)^2}$$

Notice that this is never zero if $ad - bc \neq 0$.

In Solved Exercise 2, we saw that $x = g(y) = (-dy + b)/(cy - a)$, so

$$g'(y) = \frac{-d(cy - a) - c(-dy + b)}{(cy - a)^2} = \frac{ad - bc}{(cy - a)^2}$$

As they stand, $f'(x) = (ad - bc)/(cx + d)^2$ and $g'(y) = (ad - bc)/(cy - a)^2$ do not appear to be reciprocals of one another. However, the inverse function theorem says that $g'(y) = 1/[f'(g(y))]$, so we must substitute $g(y) = (-dy + b)/(cy - a)$ for x in $f'(x)$. We obtain

$$f'(g(y)) = \cfrac{ad - bc}{\left[c\left(\cfrac{-dy + b}{cy - a} \right) + d \right]^2}$$

$$= \cfrac{ad - bc}{\left[\cfrac{-cdy + cb + cdy - ad}{cy - a} \right]^2}$$

$$= \cfrac{ad - bc}{\left[\cfrac{bc - ad}{cy - a} \right]^2} = \cfrac{(cy - a)^2}{ad - bc}$$

which is $1/[g'(y)]$.

11. Since f is a polynomial, it is continuous. The derivative is $f'(x) = 3x^2 + 2$. This is positive for $x > 0$ and in particular for x in $(0,2)$. By Theorem 1, f has an inverse function $g(y)$. We notice that $g(4) = 1$, since $f(1) = 4$. From Theorem 2, with $y_0 = 4$ and $x_0 = 1$,

$$g'(4) = \frac{1}{f'(1)} = \frac{1}{3 \cdot 1^2 + 2} = \frac{1}{5}$$

12. Let $f(x) = x^3$. Then f has inverse $g(y) = y^{1/3}$. $f'(0) = 0$, but g is not differentiable at $y = 0$. We can prove g is not differentiable at $y = 0$ as follows: we shall show that no line $l(y) = my$ through $(0,0)$ overtakes $g(y)$ at $y = 0$. Indeed, suppose it did; then in particular, in some interval J about 0, $y \in J$, $y > 0$ implies $my > y^{1/3}$, i.e., $m > 0$ and $y^{2/3} > 1/m$. But this cannot be, for it implies $y > 1/m^{3/2} > 0$ and the inequality was supposed to hold for all $y \in J$, $y > 0$, not merely those which are greater than $1/m^{3/2}$.

13. Since $g'(y_0) = 1/f'(x_0)$, g' is either > 0 or < 0 in J, as this is true for f' on I. Hence we know Theorem 2 applies with f replaced by g. The inverse of g is nothing but f. We may conclude, by Theorem 2, that f is differentiable at $x_0 = g(y_0)$ and

$$f'(x_0) = \frac{1}{g'(y_0)}$$

But

$$\frac{1}{g'(y_0)} = \frac{1}{1/f'(x_0)} = f'(x_0)$$

so this just reduces to something we already knew.

14. We find $(f \circ g)(x) = f(g(x)) = f(x + 1) = (x + 1)^2$. (Or $(f \circ g)(x) = f(u) = u^2 = (x + 1)^2$.) On the other hand, $(g \circ f)(x) = g(f(x)) = g(x^2) = x^2 + 1$. Note that $g \circ f \neq f \circ g$.

15. We observe that $h(x)$ can be expressed in terms of x^{12} as $(x^{12})^2 + 3(x^{12}) + 1$. If $u = g(x) = x^{12}$, then $h(x) = f(g(x))$, where $f(u) = u^2 + 3u + 1$.

16. (a) $D_f = (-\infty, \infty)$, $D_g = [0, \infty)$.

(b) The domain of $f \circ g$ consists of all x in D_g for which $g(x) \in D_f$; that is, all x in $[0, \infty)$ for which $\sqrt{x} \in (-\infty, \infty)$. Since $\sqrt{x}$ is always in $(-\infty, \infty)$ the domain of $f \circ g$ is $[0, \infty)$. We have $(f \circ g)(x) = f(g(x)) = f(\sqrt{x}) = \sqrt{x} - 1$.

The domain of $g \circ f$ consists of all x in D_f for which $f(x) \in D_g$; that is, all x in $(-\infty, \infty)$ for which $x - 1 \in [0, \infty)$. But $x - 1 \in [0, \infty)$ means $x - 1 \geqslant 0$, or $x \geqslant 1$, so the domain of $g \circ f$ is $[1, \infty)$. We have $(g \circ f)(x) = g(x - 1) = \sqrt{x - 1}$.

(c) $(f \circ g)(2) = \sqrt{2} - 1$; $(g \circ f)(2) = \sqrt{2 - 1} = \sqrt{1} = 1$.

(d) See Fig. S-8-4.

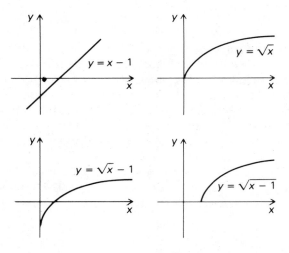

Fig. S-8-4 Graphs of $x - 1$, $\sqrt{x}$, and their compositions.

17. $(i \circ f)(x) = i(f(x)) = f(x)$, so $i \circ f = f$; $(f \circ i)(x) = f(i(x)) = f(x)$, so $f \circ i = f$.

18. The function $1/g(x)$ is the composite of $f(u) = 1/u$ and $g(x)$, i.e., $1/g(x) = f(g(x))$. Hence the derivative is, by the chain rule, $f'(g(x)) \cdot g'(x) = (1/u^2) \cdot g'(x)$, where $u = g(x)$ i.e., $(d/dx)[1/g(x)] = -g'(x)/g(x)^2$, the rule we know from Chapter 3.

19. Let D be the distance from y_0 to the nearest endpoint of J. From the inequality $|g(x) - g(x_0)| < M|x - x_0|$ (see Fig. 8-8), we see that I should be chosen small enough so that $|x - x_0| < D/M$ for all x in I.

CHAPTER 9

1. (a) By the product rule,

$$\frac{d}{dx} \sin x \cos x = \cos x \cos x + \sin x (-\sin x)$$

$$= \cos^2 x - \sin^2 x$$

(The answer $\cos 2x$ is also acceptable.)

(b) By the quotient and chain rules,
$$\frac{d}{dx} \frac{\tan 3x}{1 + \sin^2 x} = \frac{(3 \sec^2 3x)(1 + \sin^2 x) - (\tan 3x)(2 \sin x \cos x)}{(1 + \sin^2 x)^2}$$

(c)

$$\frac{d}{dx} (1 - \csc^2 5x) = -2 \csc 5x \frac{d}{dx} (\csc 5x)$$

$$= 10 \cos^2 5x \cot 5x$$

2. By the chain rule,

$$\frac{d}{d\theta} \sin(\sqrt{3\theta^2 + 1}) = \cos(\sqrt{3\theta^2 + 1}) \frac{d}{d\theta} \sqrt{3\theta^2 + 1}$$

$$= \cos(\sqrt{3\theta^2 + 1}) \cdot \frac{1}{2} \frac{3 \cdot 2\theta}{\sqrt{3\theta^2 + 1}}$$

$$= \frac{3\theta}{\sqrt{3\theta^2 + 1}} \cos\sqrt{3\theta^2 + 1}$$

3. If $f(x) = \sin^2 x$, $f'(x) = 2 \sin x \cos x$ and $f''(x) = 2(\cos^2 x - \sin^2 x)$. The first derivative vanishes when either $\sin x = 0$ or $\cos x = 0$, at which points f'' is positive and negative, yielding minima and maxima. Thus the minima of f are at $0, \pm\pi, \pm 2\pi, \ldots$, where $f = 0$, and the maxima are at $\pm\pi/2, \pm 3\pi/2, \ldots$, where $f = 1$.

The function $f(x)$ is concave up when $f''(x) > 0$ (that is, $\cos^2 x > \sin^2 x$) and down when $f''(x) < 0$ (that is, $\cos^2 x < \sin^2 x$). Also, $\cos x = \pm \sin x$ exactly if $x = \pm\pi/4$, $\pm\pi/4 \pm \pi$, $\pm\pi/4 \pm 2\pi$, and so on (see the graphs of sine and cosine). These are then inflection points separating regions of concavity and convexity. The graph is shown in Fig. S-9-1.

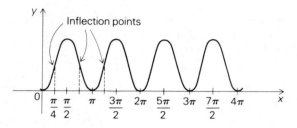

Fig. S-9-1 The graph $y = \sin^2 x$.

4. (a) Since $\sin(\pi/6) = \frac{1}{2}$, $\sin^{-1}(\frac{1}{2}) = \pi/6$. Similarly, $\sin^{-1}(\sqrt{3}/2 = -\pi/3$. Finally, $\sin^{-1}(2)$ is not defined since 2 is not in the domain $[-1,1]$ of $\sin^{-1}$.

   (b) From Fig. S-9-2 we see that $\theta = \sin^{-1} x$ (that is, $\sin\theta = |AB|/|OB| = x$) and $\tan\theta = x/\sqrt{1-x^2}$, so $\tan(\sin^{-1} x) = x/\sqrt{1-x^2}$.

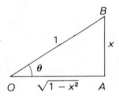

Fig. S-9-2 Finding $\tan(\sin^{-1} x)$.

5. By the chain rule,

$$\frac{d}{dx}(\sin^{-1} 2x)^{3/2} = \frac{3}{2}(\sin^{-1} 2x)^{1/2}\frac{d}{dx}\sin^{-1} 2x$$

$$= \frac{3}{2}(\sin^{-1} 2x)^{1/2}\cdot 2\cdot\frac{1}{\sqrt{1-(2x)^2}}$$

$$= 3\left(\frac{\sin^{-1} 2x}{1-4x^2}\right)^{1/2}$$

6. By the chain rule,

$$\frac{d}{dx}\sin^{-1}\sqrt{1-x^2} = \frac{1}{\sqrt{1-(\sqrt{1-x^2})^2}}\cdot\frac{d}{dx}\sqrt{1-x^2}$$

$$= \frac{1}{\sqrt{x^2}}\cdot\frac{-x}{\sqrt{1-x^2}}$$

$$= \frac{-1}{\sqrt{1-x^2}}$$

This answer is the same as the derivative of $\cos^{-1} x$. In fact,

$$\sin^{-1}\sqrt{1-x^2} = \cos^{-1} x$$

CHAPTER 10

1. $8^{-2/3} = 1/8^{2/3} = 1/(\sqrt[3]{8})^2 = 1/2^2 = 1/4$ and $8^{1/2} = \sqrt{8} \approx 2.8$

2. $9^{3/2} = (\sqrt{9})^3 = 3^3 = 27$

3. $\dfrac{(x^{2/3})^{5/2}}{x^{1/4}} = x^{(2/3)\cdot(5/2) - (1/4)} = x^{(5/3) - (1/4)} = x^{17/12}$

4. If $q = 0$, $(b^p)^q = (b^p)^0 = 1$ since $c^0 = 1$ for any c. Also, $b^{pq} = b^{p\cdot 0} = b^0 = 1$, so $(b^p)^q = b^{pq}$ in case $q = 0$. If $p = 0$, $(b^p)^q = 1^q = 1$ (since all roots and powers of 1 are 1) and $b^{pq} = b^0 = 1$.

5. It is the reflection in the y axis, since $\exp_{1/b}x = (1/b)^x = (b^{-1})^x = b^{-x} = \exp_b(-x)$.

6. $(2^{\sqrt{3}} + 2^{-\sqrt{3}})(2^{\sqrt{3}} - 2^{-\sqrt{3}}) = (2^{\sqrt{3}})^2 - (2^{-\sqrt{3}})^2 = 2^{2\sqrt{3}} - 2^{-2\sqrt{3}}$

7. (a)–(C), (b)–(B), (c)–(D), (d)–(A).

8. $\log_2 4 = 2$ since $2^2 = 4$, $\log_3 81 = 4$ since $3^4 = 81$, and $\log_{10} 0.01 = -2$ since $10^{-2} = 0.01$.

9. (a) $\log_b (b^{2x}/2b) = \log_b (b^{2x-1}/2) = \log_b (b^{2x-1}) - \log_b 2 = (2x - 1) - \log_b 2$.

   (b) $\log_2 x = \log_2 5 + 3\log_2 3 = \log_2 5\cdot 3^3$, so $x = 5\cdot 3^3 = 135$, since the function $y = \log_b x$ is one-to-one (i.e., to each value of y, there corresponds one and only one value of x).

10. (a) Suppose $x_1 < x_2$ and $0 < \lambda < 1$. Then strict convexity would mean
    $$(\lambda x_1 + (1 - \lambda)x_2)^2 < \lambda x_1^2 + (1 - \lambda)x_2^2$$
    To see this, look at the right-hand side minus the left-hand side:
    $$\begin{aligned}(\lambda x_1^2 &+ (1 - \lambda)x_2^2) - (\lambda x_1 + (1 - \lambda)x_2)^2 \\ &= (\lambda - \lambda^2)x_1^2 + ((1 - \lambda) - (1 - \lambda)^2)x_2^2 - 2\lambda(1 - \lambda)x_1 x_2 \\ &= (\lambda(1 - \lambda))(x_1^2 + x_2^2 - 2x_1 x_2) \\ &= \lambda(1 - \lambda)(x_1 - x_2)^2\end{aligned}$$
    This is strictly positive.

    (b) A constant function; e.g., $f(x) = 0$.

11. Write $x_2 = \lambda x_1 + (1 - \lambda)x_3$, where $\lambda = (x_3 - x_2)/(x_3 - x_1)$. By convexity,

$$f(x_2) \leqslant \lambda f(x_1) + (1 - \lambda)f(x_3)$$

$$= \frac{x_3 - x_2}{x_3 - x_1} f(x_1) + \left(1 - \frac{x_3 - x_2}{x_3 - x_1}\right)f(x_3)$$

Hence $(x_3 - x_2)f(x_1) \geqslant (x_3 - x_1)f(x_2) - ((x_3 - x_1) - (x_3 - x_2))f(x_3)$, which simplifies to the desired inequality. If f is strictly convex, $\geqslant$ can be replaced by $>$. The inequality obtained by isolating $f(x_3)$ instead of $f(x_1)$ is

$$f(x_3) \geqslant f(x_2) + \frac{f(x_2) - f(x_1)}{x_2 - x_1}(x_3 - x_2)$$

These inequalities are illustrated in Fig. S-10-1.

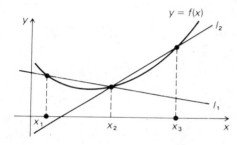

Fig. S-10-1 $f(x_3) \geqslant l_1(x_3)$ and $f(x_1) \geqslant l_2(x_1)$.

12. Let $a < x_1 < x_2 \leqslant b$. Using the inequality from Solved Exercise 11,

$$f(x_2) \geqslant f(x_1) + \frac{f(x_1) - f(a)}{x_1 - a}(x_2 - x_1) > f(x_1)$$

since

$$\frac{f(x_1) - f(a)}{x_1 - a}(x_2 - x_1) > 0.$$

Thus $f(x_2) > f(x_1)$ so f is increasing. (If $x_1 = a$, we are given $f(x_2) > f(x_1)$ by hypothesis.)

13. If we set $y = 1$ we find that $f(x) = (f(1))^x$, so merely choose $b = f(1)$. Notice that by the laws of exponents, b^x does satisfy $f(xy) = [f(y)]^x$.

14. (a) $2e^{2x}$ (b) $(\ln 2)2^x$ (c) $3xe^{3x} + e^{3x}$

(d) $(2x + 2)\exp(x^2 + 2x)$ (e) $2x$

15. We have $\exp_b'(0) = \ln b$, so

$$\exp_b\left(\frac{1}{\exp_b'(0)}\right) = \exp_b\left(\frac{1}{\ln b}\right) = b^{1/\ln b} = (e^{\ln b})^{1/\ln b} = e^{\ln b(1/\ln b)} = e^1 = e$$

16. (a) $e^{\sqrt{x}}(1/2\sqrt{x})$ by the chain rule.

   (b) $e^{\sin x} \cdot \cos x$

   (c) By the chain rule, $(d/dx)b^{u(x)} = b^{u(x)}\ln b \cdot (du/dx)$. Here $(d/dx)2^{\sin x} = 2^{\sin x} \cdot \ln 2 \cdot \cos x$.

   (d) $2\sin x \cos x$

17. By the mean value theorem, and letting $f(t) = b^t$,

$$b^t - 1 = f(t) - f(0)$$
$$= f'(c)(t - 0)$$

for some c between 0 and t. But $f'(c) = b^c \log_e b < b^t \log_e b$ since b^t is increasing and $\log_e b > 0$. Hence

$$b^t - 1 < (b^t \log_e b)t$$

18. (a) By the chain rule, $(d/dx)\ln 10x = [(d/du)\ln u](du/dx)$, where $u = 10x$; that is, $(d/dx)\ln 10x = 10 \cdot 1/10x = 1/x$.

   (b) $(d/dx)\ln u(x) = u'(x)/u(x)$ by the chain rule.

   (c) $\cos x/\sin x = \cot x$ (We use the preceding solution.)

   (d) Use the product rule: $(d/dx)\sin x \ln x = \cos x \ln x + (\sin x) \cdot 1/x$.

   (e) Use the quotient rule: $(d/dx)\ln x/x = (x \cdot 1/x - \ln x)/x^2 = (1 - \ln x)/x^2$.

   (f) By the differentiation formula on p. 143: $(d/dx)\log_5 x = 1/(\ln 5)x$.

19. (a) Write $x^n = e^{(\ln x)n}$ and differentiate using the chain rule:

$$\frac{d}{dx}x^n = \frac{d}{dx}e^{(\ln x) \cdot n} = \frac{n}{x} \cdot e^{(\ln x)n}$$

$$= \frac{n}{x}x^n = nx^{n-1}$$

   using the laws of exponents.

   (b) $(d/dx)x^\pi = \pi x^{\pi-1}$, by part (a).

20. $\ln y = \ln[(2x + 3)^{3/2}/(x^2 + 1)^{1/2}] = \frac{3}{2}\ln(2x + 3) - \frac{1}{2}\ln(x^2 + 1)$, so

$$\frac{1}{y}\frac{dy}{dx} = \frac{3}{2} \cdot \frac{2}{2x + 3} - \frac{1}{2} \cdot \frac{2x}{x^2 + 1} = \frac{3}{2x + 3} - \frac{x}{x^2 + 1} = \frac{(x^2 - 3x + 3)}{(2x + 3)(x^2 + 1)}$$

and hence

$$\frac{dy}{dx} = \frac{(2x + 3)^{3/2}}{(x^2 + 2)^{1/2}} \cdot \frac{(x^2 - 3x + 3)}{(2x + 3)(x^2 + 1)} = \frac{(x^2 - 3x + 3)(2x + 3)^{1/2}}{(x^2 + 1)^{3/2}}$$

21. We find that $\ln y = x^x \ln x$, so using the fact that $(d/dx)x^x = x^x(1 + \ln x)$ from Worked Example, we get

$$\frac{1}{y}\frac{dy}{dx} = x^x(1 + \ln x)\ln x + \frac{x^x}{x}$$

so

$$\frac{dy}{dx} = x^{(x^x)}[x^x(1 + \ln x)\ln x + x^{x-1}]$$

CHAPTER 11

1. The partitions are shown in Fig. S-11-1.

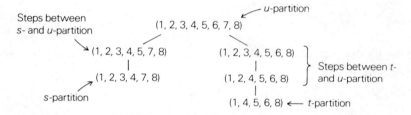

Fig. S-11-1 Connecting two partitions.

2. (a) If $0 < x \leqslant 1$, we can take $(0,x)$ as our adapted partition, and

$$\int_0^x f(t)dt = 2(x - 0) = 2x$$

If $1 < x \leqslant 3$, we take $(0,1,x)$ as our adapted partition, and

$$\int_0^x f(t)dt = 2 \cdot (1 - 0) + 0 \cdot (x - 1) = 2$$

If $3 < x \leqslant 4$, we take $(0,1,3,x)$ as our adapted partition, and

$$\int_0^x f(t)dt = 2(1 - 0) + 0(3 - 1) + (-1)(x - 3)$$

$$= 2 - x + 3$$
$$= -x + 5$$

Summarizing, we have

$$\int_0^x f(t)dt = \begin{cases} 2x & \text{if } 0 < x \leqslant 1 \\ 2 & \text{if } 1 < x \leqslant 3 \\ -x + 5 & \text{if } 3 < x \leqslant 4 \end{cases}$$

(b) The graph of $F(x)$ is shown in Fig. S-11-2. Functions like F are sometimes called *piecewise linear, polygonal,* or *ramp* functions.

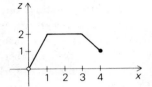

Fig. S-11-2 A piecewise
linear function.

(c) We can see from the graph that F is differentiable everywhere on $[0,4]$
except at the points 0, 1, 3, and 4. We have:

$$F'(x) = \begin{cases} 2 & \text{if } 0 < x < 1 \\ 0 & \text{if } 1 < x < 3 \\ -1 & \text{if } 3 < x < 4 \end{cases}$$

Comparing this with the definition of $f(t)$ on p. 153, we see that F' is
the same as f, except at the points where the piecewise constant function f has a jump. This example illustrates the general fact that the
derivative of the integral of a function, with respect to an endpoint of
integration, is more or less equal to the original function. The fundamental theorem of calculus is just this fact, made precise and extended
to functions which are not piecewise constant.

3. By comparing f with the constant functions 0 and 1, we see first of all that
0 is a lower sum and 1 is an upper sum. It follows from Theorem 2 that L_f
contains the interval $(-\infty, 0]$ and U_f contains $[1, \infty)$. We will now show that
the interval $(0,1)$ is a gap between L_f and U_f; i.e., the numbers in $(0,1)$ are
neither upper nor lower sums.
    Let g be a piecewise constant function such that $g(t) \leq f(t)$ for all t in
(a, b), $(t_0, t_2, \dots, t_n)$ an adapted partition, and $g(t) = k_i$ for t in (t_{i-1}, t_i).
Since the interval (t_{i-1}, t_i) contains some irrational numbers (see Solved
Exercise 13, Chapter 4), where $f(t) = 0$, it follows that $k_i \leq 0$. Then $k_i \Delta t_i \leq$
0, and the lower sum $\Sigma_{i-1}^n (k_i \Delta t_i)$ is less than or equal to zero. We have thus
shown that every lower sum lies in $(-\infty, 0]$. Similarly (using the fact that
every open interval contains some rational numbers), we can show that
every upper sum lies in $[1, \infty)$.
    The function f may seem rather bizarre, but it is a useful mathematical
example, which reminds us that some conditions are necessary in order to
form integrals. Specifically, this example shows that it is a mistake to
assume that every bounded function is integrable.

4. Let f be piecewise constant on $[a, b]$. Let $g(t) = f(t)$. Then g is piecewise
constant on $[a, b]$, and $g(t) \leq f(t)$ for all t, so

$$S_0 = \int_a^b g(t)dt = \int_a^b f(t)dt$$

is a lower sum. (Here the integral signs refer to the integral as defined on p. 149.) Since every number less than a lower sum is again a lower sum, we conclude that every number less than S_0 is a lower sum. But we also have $g(t) \geqslant f(t)$ for all t, so S_0 is an upper sum as well, and so is every number greater than S_0. Thus, by the definition on p. 155, f is integrable on $[a,b]$, and its integral is S_0, which was the integral of f according to the definition on p. 149.

5. Let g_0 and h_0 be piecewise constant functions on $[a,x_0]$, with $g_0(t) \leqslant f(t) \leqslant h_0(t)$ for all t in (a,x_0), such that $\int_a^{x_0} h_0(t)dt - \int_a^{x_0} g_0(t)dt < \epsilon$. We let $g(t)$ and $h(t)$ be the restrictions of $g_0(t)$ and $h_0(t)$ to the interval $[a,x]$. To show that $\int_a^x h(t)dt - \int_a^x g(t)dt$ is less than ϵ, we choose a partition $(t_0,\ldots,t_n)$ of $[a,x_0]$ with the following properties: $(t_0,\ldots,t_n)$ is adapted to both g_0 and h_0; the point x is one of the partition points, say, $x = t_m$. (We obtain such a partition by putting together partitions for g_0 and h_0, throwing in the point x, eliminating repetitions, and putting all the points in order.) Now the partition $(t_0,\ldots,t_m)$ is adapted to g and h, and we have the following formulas for the integrals, where k_i is the value of g on (t_{i-1},t_i) and l_i is the value of h on (t_{i-1},t_i).

$$\int_a^x g(t)dt = \sum_{i=1}^m k_i \Delta t_i, \qquad \int_a^{x_0} g_0(t)dt = \sum_{i=1}^n k_i \Delta t_i$$

$$\int_a^x h(t)dt = \sum_{i=1}^m l_i \Delta t_i, \qquad \int_a^{x_0} h_0(t)dt = \sum_{i=1}^n l_i \Delta t_i$$

Subtracting in each column gives

$$\int_a^x h(t)dt - \int_a^x g(t)dt = \sum_{i=1}^m (l_i - k_i) \Delta t_i$$

$$\int_a^{x_0} h_0(t)dt - \int_a^{x_0} g_0(t)dt = \sum_{i=1}^n (l_i - k_i) \Delta t_i$$

The terms $(l_i - k_i)\Delta t_i$ are all nonnegative, because $g_0(t) \leqslant h_0(t)$ for t in $[a,x_0]$. Since the sum on the right involves all the terms of the sum on the left, plus $n - m$ extra ones, we have

$$\int_a^x h(t)dt - \int_a^x g(t)dt \leqslant \int_a^{x_0} h_0(t)dt - \int_a^{x_0} g_0(t)dt$$

The right-hand side was assumed to be less than ϵ, so the left-hand side is, too, and so f is ϵ-integrable on $[a, x]$.

6. Suppose that f is increasing on $[a, b]$; i.e., $f(x) \leqslant f(y)$ whenever $a \leqslant x \leqslant y \leqslant b$. Given a positive integer n, we may divide the integral $[a, b]$ into n subintervals of equal length. The length of each subinterval will be $(b - a)/n$, and the partition points will be $(x_0, x_1, \ldots, x_n)$, where $x_0 = a$, $x_1 = a + (b - a)/n$, $x_2 = a + 2(b - a)/n, \ldots, x_i = a + i(b - a)/n, \ldots, x_n = a + n[(b - a)/n] = a + b - a = b$. The ith interval is

$$[x_{i-1}, x_i] = \left[a + \frac{(i - 1)(b - a)}{n}, a + \frac{i(b - a)}{n} \right]$$

Since f is increasing on $[a, b]$, it follows that the lowest value of $f(x)$ for x in the ith interval is $f(x_{i-1})$; the highest value is $f(x_i)$. Let us define the piecewise constant functions g and h on $[a, b]$ by

$$g(x) = \begin{cases} f(x_{i-1}) & \text{for } x_{i-1} \leqslant x < x_i \\ f(x_{n-1}) & \text{for } x = x_n \end{cases}$$

$$h(x) = \begin{cases} f(x_i) & \text{for } x_{i-1} \leqslant x < x_i \\ f(x_n) & \text{for } x = x_n \end{cases}$$

Then, for all x in $[a, b]$, we have $g(x) \leqslant f(x) \leqslant h(x)$ (see Fig. S-11-3).

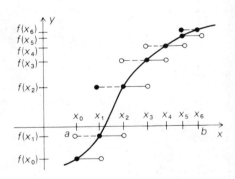

Fig. S-11-3 The graph of $f(x)$ (curved line) lies above the graph of $g(x)$ (solid horizontal segments) and below the graph of $h(x)$ (dashed horizontal segments).

Since the integral of f is a transition point between its upper and lower sums, we must have

$$\int_a^b g(x)\,dx \leqslant \int_a^b f(x)\,dx \leqslant \int_a^b h(x)\,dx$$

Now we may calculate the upper and lower sums explicitly, since we have formulas for $g(x)$ and $h(x)$. Using the fact that $\Delta x_i = x_i - x_{i-1} = (b-a)/n$ for all i, we have

$$L_n = \int_a^b g(x)dx = f(x_0)\frac{b-a}{n} + f(x_1)\frac{b-a}{n} + \cdots + f(x_{n-1})\frac{b-a}{n}$$

$$= [f(x_0) + f(x_1) + \cdots + f(x_{n-1})]\frac{b-a}{n}$$

$$= \frac{b-a}{n}\sum_{i=0}^{n-1} f(x_i)$$

and

$$U_n = \int_a^b h(x)dx = f(x_1)\frac{b-a}{n} + f(x_2)\frac{b-a}{n} + \cdots + f(x_n)\frac{b-a}{n}$$

$$= [f(x_1) + f(x_2) + \cdots + f(x_n)]\frac{b-a}{n}$$

$$= \frac{b-a}{n}\sum_{i=1}^{n} f(x_i)$$

so we may conclude that

$$L_n = \frac{b-a}{n}\sum_{i=0}^{n-1} f(x_i) \leqslant \int_a^b f(x)dx \leqslant \frac{b-a}{n}\sum_{i=1}^{n} f(x_i) = U_n$$

where $x_i = a + i(b-a)/n$.

Now we note that

$$U_n - L_n = \frac{b-a}{n}\sum_{i=1}^{n} f(x_i) - \frac{b-a}{n}\sum_{i=0}^{n-1} f(x_i)$$

$$= \frac{b-a}{n}[f(x_n) - f(x_0)] = \frac{b-a}{n}[f(b) - f(a)]$$

so if we have computed L_n, we can obtain U_n simply by adding to it $[(b-a)/n][f(b) - f(a)]$.

The last calculation shows that the difference $U_n - L_n$ between the upper and lower sums becomes smaller and smaller as n becomes larger and larger. In fact, given any $\epsilon > 0$, $[(b-a)/n][f(b) - f(a)] < \epsilon$ if n is chosen large enough. By Problem 13, Chapter 4, there is a transition point between the upper and lower sums; f is integrable on $[a, b]$.

7. The integral exists, since $1/t$ is continuous on $[1,2]$. To estimate it to within $\frac{1}{10}$, we try to find lower and upper sums which are within $\frac{2}{10}$ of one

another. We divide the integral into three parts and use the best possible piecewise constant function, as shown in Fig. S-11-4. For a lower sum we have

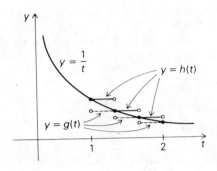

$y = \dfrac{1}{t}$

$y = h(t)$

$y = g(t)$

Fig. S-11-4 Estimating $\int_1^2 (1/t)\, dt$.

$$\int_1^2 g(t)\, dt = \frac{1}{4/3}\left(\frac{4}{3} - 1\right) + \frac{1}{5/3}\left(\frac{5}{3} - \frac{4}{3}\right) + \frac{1}{2}\left(2 - \frac{5}{3}\right)$$

$$= \frac{3}{4}\cdot\frac{1}{3} + \frac{3}{5}\cdot\frac{1}{3} + \frac{1}{2}\cdot\frac{1}{3}$$

$$= \frac{1}{3}\left(\frac{3}{4} + \frac{3}{5} + \frac{1}{2}\right)$$

$$= \frac{1}{3}\left(\frac{37}{20}\right) = \frac{37}{60}$$

For an upper sum we have

$$\int_1^2 h(t)\, dt = \frac{1}{1}\cdot\frac{1}{3} + \frac{1}{4/3}\cdot\frac{1}{3} + \frac{1}{5/3}\cdot\frac{1}{3}$$

$$= \frac{1}{3}\left(1 + \frac{3}{4} + \frac{3}{5}\right)$$

$$= \frac{1}{3}\left(\frac{47}{20}\right) = \frac{47}{60}$$

It follows that

$$\frac{37}{60} \leqslant \int_1^2 \frac{1}{t}\, dt \leqslant \frac{47}{60}$$

Since the integral lies in the interval $\left[\frac{37}{60}, \frac{47}{60}\right]$, whose length is $\frac{1}{6}$, we may take the midpoint $\frac{42}{60} = \frac{7}{10}$ as our estimate; it will differ from the true integral by no more than $\frac{1}{2}\cdot\frac{1}{6} = \frac{1}{12}$, which is less than $\frac{1}{10}$.

8. If we examine the upper and lower sums in Solved Exercise 7, we see that they differ by the "outer terms"; the difference is

$$\frac{1}{3}\left(1-\frac{1}{2}\right)=\frac{1}{3}\cdot\frac{1}{2}=\frac{1}{6}$$

If, instead of three steps, we used n steps, the difference between the upper and lower sums would be $(1/n)\cdot(1/2)$. To estimate the interval to within $\frac{1}{100}$, we would have to make $(1/2n)\leqslant\frac{1}{50}$, which we could do by putting $n=25$.

Computing these upper and lower sums (with a calculator), we obtain

$$0.6832\leqslant\int_{1}^{2}\frac{1}{t}dt\leqslant0.7032$$

so our estimate is 0.6932.

9. We use the technique of "splitting the difference." Let $\epsilon=p_1+p_2-m$, which is positive. Put $m_1=p_1-(\epsilon/2)$ and $m_2=p_2-(\epsilon/2)$. Then $m_1<p_1$, $m_2<p_2$, and $m_1+m_2=p_1+p_2-\epsilon=m$.

10. f is integrable on $[0,1]$ because it is constant there. It is tempting to apply Theorem 3 to f on $[1,2]$ but f is not continuous at 1. However, altering f at an endpoint does not change its integrability (the upper and lower sums are the same); since the function x^2 is integrable on $[1,2]$, and f agrees with it except at 1, f must be integrable on $[1,2]$. Now, by part 1 of Theorem 4, f is integrable on $[0,2]$.

11. Since $h(t)=t_i=a+[i(b-a)]/n$ on (t_{i-1},t_i) (see Fig. S-11-5), we have

$$\int_{a}^{b}h(t)dt=\sum_{i=1}^{n}k_i\Delta t_i=\sum_{i=1}^{n}\left[a+\frac{i(b-a)}{n}\right]\left(\frac{b-a}{n}\right)$$

$$=\sum_{i=1}^{n}\left[\frac{a(b-a)}{n}+\frac{i(b-a)^2}{n^2}\right]$$

$$=\sum_{i=1}^{n}\frac{a(b-a)}{n}+\frac{(b-a)^2}{n^2}\sum_{i=1}^{n}i$$

$$=n\cdot\frac{a(b-a)}{n}+\frac{(b-a)^2}{n^2}\frac{n(n+1)}{2}$$

$$=a(b-a)+\frac{(b-a)^2(n+1)}{2n}$$

$$=ab-a^2+\frac{(b-a)^2}{2}+\frac{(b-a)^2}{2n}$$

$$=\frac{b^2-a^2}{2}+\frac{(b-a)^2}{2n}$$

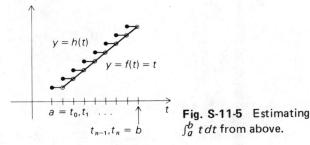

Fig. S-11-5 Estimating $\int_a^b t\,dt$ from above.

12. We repeat the calculation of $\int_a^b t\,dt$, except that we make $g(t) = 5t_{i-1}$ $\leqslant t <$ t_i and $h(t) = 5t_i$ for $t_{i-1} \leqslant t < t_i$. Then all the terms in the integrals of g and h are multiplied by 5, so

$$\int_a^b g(t)dt = \frac{5(b^2 - a^2)}{2} - \frac{5(b - a)^2}{2n}$$

and

$$\int_a^b h(t)dt = \frac{5(b^2 - a^2)}{2} + \frac{5(b - a)^2}{2n}$$

Since $g(t) \leqslant 5t \leqslant h(t)$ on $[a, b]$, we have for all n,

$$\frac{5(b^2 - a^2)}{2} - \frac{5(b - a)^2}{2n} \leqslant \int_a^b 5t\,dt$$

$$\leqslant \frac{5(b^2 - a^2)}{2} + \frac{5(b - a)^2}{2n}$$

It follows that $\int_a^b 5t\,dt = [5(b^2 - a^2)]/2$.

13. (a) The region is sketched in Fig. S-11-6.

(b) The area of the region $BCED$ is the difference between the areas of the triangles ADE and ABC; that is, $\frac{1}{2}b^2 - \frac{1}{2}a^2 = \frac{1}{2}(b^2 - a^2)$, which is equal to $\int_a^b t\,dt$.

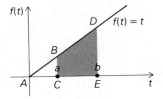

Fig. S-11-6 Area of a trapezoid as an integral.

CHAPTER 12

1. An antiderivative of x^4 is $\frac{1}{5}x^5$, since $(d/dx)(\frac{1}{5}x^5) = \frac{1}{5} \cdot 5x^4 = x^4$, so

$$\int_0^1 x^4 \, dx = \frac{1}{5}x^5 \Big|_0^1 = \frac{1}{5}(1^5) - \frac{1}{5}(0^5) = \frac{1}{5}.$$

(Notice that we use x everywhere here just as we used t before.)

2. By the sum, constant multiple, and power rules, an antiderivative for $t^2 + 3t$ is $(t^3/3) + (3t^2/2)$, and so

$$\int_0^3 (t^2 + 3t) \, dt = \left(\frac{t^3}{3} + \frac{3t^2}{2} \right) \Big|_0^3 = \frac{3^3}{3} + \frac{3 \cdot 3^2}{2} = \frac{45}{2}.$$

3. Velocity is defined as the time derivative of position; that is, $v = dx/dt$, where $x = F(t)$ is the position at time t. The fundamental theorem gives the equation

$$F(b) - F(a) = \int_a^b F'(t) \, dt = \int_a^b f(t) \, dt$$

or

$$\Delta x = \int_a^b \frac{dx}{dt} \, dt = \int_a^b v \, dt$$

So the integral of the velocity v over the time interval $[a, b]$ is the *total displacement* of the object from time a to time b.

4. By Corollary 1 to the mean value theorem, we must have

$$\frac{F(\frac{1}{3}) - F(0)}{\frac{1}{3} - 0} < 2$$

and

$$\frac{F(2) - F(\frac{1}{3})}{2 - \frac{1}{3}} < 1$$

Hence

$$F(\tfrac{1}{3}) - F(0) < \tfrac{2}{3}$$

and

$$F(2) - F(\tfrac{1}{3}) < \tfrac{5}{3}$$

Adding the last two equations gives the conclusion

$$F(2) - F(0) < \tfrac{7}{3}$$

This is a simple example of one of the basic ideas involved in the proof of the fundamental theorem of calculus.

5. We let $h(s) = \int_0^s f(t)dt$, so that $F(x) = h(g(x))$. By the chain rule, $F'(x_0) = h'(g_0)) g'(x_0)$. By Theorem 2, $h'(s) = f(s)$, so

$$F'(x_0) = f(g(x_0)) \cdot g'(x_0)$$

6. By Solved Exercise 5 (with the endpoint 0 replaced by 1), we have

$$F'(x) = \frac{1}{x^2} \cdot \frac{d}{dx} (x^2) = \frac{2}{x}$$

CHAPTER 13

1. We apply the corollary to Theorem 2, with $f(x) = \tan x$, $x_0 = \pi/4$ and $\epsilon = 0.001$. Since $\tan x$ is continuous at $\pi/4$, there is a positive number δ such that, whenever $|x - \pi/4| < \delta$, $|\tan x - \tan \pi/4| < 0.001$; that is, $|\tan x - 1| < 0.001$.

2. First of all, suppose that l is a transition point from A to B. Then there is an open interval I containing l such that if $c \in I$, $c < l$ then c lies in A and not in B and if $d \in I$, $d > l$, then d lies in B and not in A. Let $c_1 < l$ and choose c such that $c_1 < c < l$ and $c \in I$. Then $c \in A$, so there is an open interval I about x_0 such that for $x \in I$, $x \neq x_0$ we have $c < f(x)$. Since $c_1 < c$ we also have $c_1 < f(x)$. This is the statement of part 1 of condition 1 in the definition of limit. The proof of part 2 is similar, and the converse statement may be proved in the same way.

   Note. This argument and the completeness axiom show that A and B, if not empty, are of the following form (for any function):

   $A = (-\infty, \underline{S}]$ or $(-\infty, \underline{S})$

   $B = [\overline{S}, \infty)$ or $(\overline{S}, \infty)$

   The end points $\underline{S}$ and $\overline{S}$ of A and B are called the *limit inferior* and *limit superior* of f, respectively.

3. If we write Δx for $x - x_0$, then $x = x_0 + \Delta x$ and $\Delta x \to 0$ when $x \to x_0$, so

$$\lim_{x \to x_0} \frac{f(x) - f(x_0)}{x - x_0} = \lim_{\Delta x \to 0} \frac{f(x_0 + \Delta x) - f(x_0)}{\Delta x} = f'(x_0)$$

4. By Theorem 2, g is continuous at 0 if and only if $\lim_{\Delta x \to 0} g(\Delta x) = g(0)$, i.e.,

$$\lim_{\Delta x \to 0} \frac{f(x_0 + \Delta x) - f(x_0)}{\Delta x} = m_0$$

But, by Theorem 3, this is true if and only if $f'(x_0) = m_0$.

5. (a) Directly,

$$S_n = \frac{1}{n^2} \sum_{i=1}^{n} (n + i)$$

$$= \frac{1}{n^2} \left[n^2 + \frac{n(n + 1)}{2} \right]$$

$$= \left(1 + \frac{1}{2} + \frac{1}{2n} \right) \to \frac{3}{2}$$

as $n \to \infty$. (We used $\sum_{i=1}^{n} i = n(n + 1)/2$.)

(b) $S_n = \sum_{i=1}^{n} (1 + i/n)/n$ is the Riemann sum for $f(x) = 1 + x$ with $a = 0$, $b = 1$, $t_0 = 0$, $t_1 = 1/n$, $t_2 = 2/n, \ldots, t_n = 1$, width $\Delta t_i = 1/n$ and $c_i = t_i$. Hence

$$S_n \to \int_0^1 (1 + x) \, dx = \frac{(1 + x)^2}{2} \Big|_0^1 = \frac{3}{2}$$

as $n \to \infty$.

6. Let $t_0, t_1, \ldots, t_n$ be a partition of $[a, b]$ with maximum width $\to 0$. Let c_i belong to $[t_{i-1}, t_i]$. Then

$$\int_a^b f(x) \, dx = \lim_{n \to \infty} \sum_{i=1}^{n} f(c_i) \, \Delta t_i$$

$$\int_a^b g(x) \, dx = \lim_{n \to \infty} \sum_{i=1}^{n} g(c_i) \, \Delta t_i$$

and

$$\int_a^b [f(x) + g(x)] \, dx = \lim_{n \to \infty} \sum_{i=1}^{n} [f(c_i) + g(c_i)] \, \Delta t_i$$

But, by the summation rules and the assumed limit law, this last expression equals

$$\lim_{n \to \infty} \sum_{i=1}^{n} f(c_i) \, \Delta t_i + \lim_{n \to \infty} \sum_{i=1}^{n} g(c_i) \, \Delta t_i.$$

$$= \int_a^b f(x) \, dx + \int_a^b g(x) \, dx$$

Index